T0134037

Compendium of Civil Engineering Education Strategies

Compendium of Civil Engineering Education Strategies

Case Studies and Examples

Hudson Jackson and Kassim Tarhini

CRC Press
Taylor & Francis Group
Boca Raton London New York

CRC Press is an imprint of the
Taylor & Francis Group, an **Informa** business

First edition published 2022
by CRC Press
6000 Broken Sound Parkway NW, Suite 300, Boca Raton, FL 33487-2742

and by CRC Press
2 Park Square, Milton Park, Abingdon, Oxon, OX14 4RN

ISBN: 978-1-032-23538-7 (hbk)
ISBN: 978-1-032-24777-9 (pbk)
ISBN: 978-1-003-28005-7 (ebk)

DOI: 10.1201/9781003280057

Typeset in Times
by codeMantra

To Jinah Jackson, Nahisha Jackson,
Maria Alexandra Lopez and Nadia Tarhini

Contents

Preface

There is an ongoing worldwide need for undergraduate engineering education to provide adequate balance between theory and practice as well as emphasize industrial relevance and global perspectives in civil engineering. Global interconnectedness now requires engineers to not only be technically competent but to also embrace diversity, be sensitive to cultural needs, be ethical, and demonstrate good leadership qualities. More than ever, engineers must be able to function in this increasingly global environment with adequate awareness of the engineering needs, practices and manage projects with the appropriate demonstration of professional and leadership skills. Furthermore, engineers should also be able to address the uncertainties associated with changing climatic conditions. Due to these competing demands, the civil engineering community is faced with the challenge of not only ensuring that infrastructures are resilient but also understand the associated impact to society, the environment, and the costs of doing business. With sound technical knowledge and understanding of cultural, environmental, business, and ethical forces within the global context, engineering graduates will be better prepared to work effectively in addressing the challenges within the changing global environment.

This book is structured in the form of a compendium of key topics in civil engineering education. One of the main objectives in writing this book is to compile into a single volume information that is relevant to civil engineering education. The book covers topics such as global perspectives, critical and design thinking skills, leadership skills, assessment, recruitment, and retention. The authors discuss general approaches used in academia to develop engineering leaders including strategies that have been successfully used in the Civil Engineering (CE) Program at the United States Coast Guard Academy (USCGA). This compendium highlights strategies that were infused into the CE curriculum without expansion or addition of new courses. Examples of curricular and co-curricular activities that provide students opportunities that help them to become critical thinkers, information literate, resilient leaders, and see the world through the eyes of others are included. The intended audiences are engineering faculty, practitioners, and graduate students considering a career in academia. Each chapter includes an overview of the academic approaches for each topic including implementation examples. The chapter on assessment includes examples of how assessment data can be used for holistic evaluation, and improvement of student learning. It is hoped that each chapter can be used separately or in combination with other chapters to help enhance and foster student learning as well as promote the development of skills required for engineering practice. Academic faculty and even practitioners will find the contents helpful as instructional and reference material in developing and assessing professional skills. The content will also be useful to intellectually curious students who want a deeper understanding and appreciation of the need for professional development and life-long learning.

The book is organized into seven chapters:

- Chapter 1 – *Global Perspective in Civil Engineering Education*: Trends in global processes affect many areas of human life including education for professional practice in all disciplines. Understanding the effects of globalization on both teaching and learning is essential to educate professionals, who would become valuable contributors to the integrated and interdependent global environment. This chapter highlights the importance of engineering students to develop a global perspective, and successful strategies that have been used or considered in educating globally minded and resilient engineers are discussed.
- Chapter 2 – *Critical and Design Thinking across the Civil Engineering Curriculum*: Engineering undergraduate programs are required to meet accreditation criteria and help students develop and hone technical and professional skills, especially critical and design thinking, to be successful professional engineers. In this chapter, strategies and a process of infusing critical and design thinking in engineering courses along with assessment tools to evaluate students' progressive development are discussed in connection with the six cognitive levels of Bloom's Taxonomy.
- Chapter 3 – *Development of Leadership Skills*: To become effective global civil engineers, students need to understand and develop skills in both leadership and management. In this chapter, the development of leadership skills including several leadership models used in academia is discussed.
- Chapter 4 – *Communication and Information Literacy Skills*: It is important for students to know that they need to be able to communicate to different audiences and that it is crucial to learn how to achieve the appropriate tone and style for the intended audience. Developing effective communication skills is essential for success in the workforce. With the vast and easy access to information, students are also expected to develop an appropriate level of "information literacy" so they can manage these vast sources of information and appropriately use them. An overview of strategies for developing communication and information literacy skills is also presented in this chapter.
- Chapter 5 – *Professional Ethics*: It has been challenging for most engineering faculty to teach ethics because of a lack of professional background to adequately address the complexities of moral and professional ethics. In addressing ethics in the engineering curriculum, a clear distinction must be made between moral behavior and professional ethics within different cultures and professional settings. To accomplish this, the curricula of engineering programs should be infused with a more global perspective on ethics and its relevance in the practice of engineering throughout the world. In this chapter, the importance of professional ethics is discussed and examples on how ethics is addressed at the undergraduate level are provided.

- Chapter 6 – *Assessment of Student Learning*: This chapter focuses on the assessment process for evaluating student learning and achievement of objectives and outcomes. It includes a general discussion of key factors required for effective assessment. Several examples of assessment tools and strategies for holistic analysis of the data are presented.
- Chapter 7 – *Engineering Outreach – Recruitment and Retention*: To address the ongoing challenges in attracting and retaining high school and first-year college students in Science, Technology, Engineering, and Mathematics, the authors discuss strategies that could be used to improve recruitment and retention in engineering.

Acknowledgments

We acknowledge our colleagues at the United States Coast Guard Academy who constantly challenge and encourage students to develop their intellectual curiosity and foster a psychologically safe space and an environment for students to grow as individuals. The support and contributions of the faculty in the Civil Engineering Department, especially Dr. Zelmanowitz, CAPT Fleischmann, CDR Maggi, and Dr. Zapalska in the Management Department, were vital in the completion of this project. We acknowledge Ms. Melissa Fernandez for her work included in Appendices 7 and 8. The extensive work that CAPT Jonathan Russell (Retired) did in developing and promoting the CG AROW program as an outreach and recruiting tool is acknowledged. We also acknowledge the contribution of Dr. Evelyn Ellis in chapter 7.

We also extend our thanks and appreciation to our families for their continued support and encouragement.

Hudson Jackson
Kassim Tarhini

Authors

Dr. Hudson Jackson is a licensed Professional Engineer with more than 30 years of consulting, academic, and research experience in geotechnical engineering, structural design, pavement engineering, construction management, field inspections, and construction quality control. His professional experience spans the continents of Africa, Europe, and North America. Dr. Jackson is currently Professor of Civil Engineering at the Coast Guard Academy in New London, Connecticut. In this role, he actively participates in the intellectual and professional development of leaders of character within the Coast Guard. He teaches courses in Coastal Resiliency, Transportation Engineering, Soil Mechanics, Geotechnical Engineering Design, and Civil Engineering Materials and is an active advisor for senior capstone projects. With a background in civil engineering consulting, Dr. Jackson also serves as a general advisor and mentor on matters of professional practice issues. Prior to his current appointment, Dr. Jackson worked as a consulting engineer at Stantec Consulting where he served as an Associate, Office Leader, Project Manager, and Field Operations Manager for several state and privately funded projects throughout the United States. Dr. Jackson was also a consulting engineer in Germany and Sierra Leone, West Africa, where he supervised several mid- to high-level structural design, highway design and construction, mining, and dam projects.

Dr. Jackson earned a doctorate (PhD) and a Master of Philosophy (Mpl) degree in Geotechnical Engineering from Rutgers University, Piscataway, New Jersey. He also earned a Master of Engineering degree in Construction and Geotechnical Engineering from the Technical Engineering University in Darmstadt, Germany. He is a licensed Professional Engineer in the United States, Germany, and Sierra Leone. Dr. Jackson has a strong passion to prepare students for the global community, which includes educating students to become better environmental stewards, and to encourage minority students to pursue STEM careers through opportunities for exposure to engineering. He has published several papers to this effect. He is active in several professional organizations and has published more than 70 papers at national and international conferences. He is a member of the American Society of Civil Engineers, Deep Foundation Institute, Geo-Institute, and the American Society for Engineering Education. Dr. Jackson grew up in Freetown, Sierra Leone, West Africa, and graduated from the University of Sierra Leone with a Bachelor of Engineering degree in Civil Engineering with Honors.

 **Dr. Kassim Tarhini** is a licensed Professional Engineer with more than 35 years of academic, research, and consulting experience in structural and geotechnical engineering. Dr. Tarhini is currently Professor of Civil Engineering at the Coast Guard Academy in New London, Connecticut. In this role, he actively contributes to the professional development of leaders of character within the Coast Guard. Currently, Dr. Tarhini teaches courses in Engineering Mechanics – Statics, Mechanics of Materials, Transportation Engineering, Reinforced Concrete Design, and occasionally Civil Engineering Materials, Soil Mechanics, and Structures. Dr. Tarhini also serves as an advisor for senior capstone projects.

Dr. Tarhini is an active member in several professional organizations such as the American Society of Civil Engineers, Society for Experimental Mechanics, and the American Society for Engineering Education. Dr. Tarhini grew up in Lebanon and pursued higher education in the United States by earning a PhD in Engineering Mechanics, MSCE (Structures and Geotechnical), and BSCE (Magna Cum Laude) from the University of Toledo, Ohio. He is passionate about engineering education and works hard to develop global citizens. Dr. Tarhini has published more than 150 papers as a result of collaboration with colleagues at the U.S. Coast Guard Academy, American University of Beirut, Valparaiso University, University of Toledo, and the University of Nevada–Las Vegas. These papers covered technical subjects and innovative pedagogy in engineering education. He collaborated with colleagues at USCGA to introduce the progressive integration of design process into the curriculum, infusing ethics activities through case studies, and the enhancement of leadership development in undergraduate engineering education. Dr. Tarhini is active in the ABET assessment process and engineering accreditation that target the development of professional responsibilities and attitudes such as leadership, communications, teamwork, ethics, critical thinking, life-long learning, cultural awareness, and globalization.

1 Global Perspectives in Civil Engineering Education

1.1 INTRODUCTION

Globalization is a multi-faceted process viewed as the ongoing cultural, social, political, economic, environmental, and technological integration of the world. The degree of its impact varies across countries, sectors of the economy, and professional communities. An important dimension of globalization is the continuous sharing of advanced technology and the spread of new knowledge, ideas, concepts, attitudes, and values. Within the context of a global market, most engineers would be required to work on international projects in countries with different languages, cultures, and engineering standards. The COVID-19 pandemic demonstrated how interconnected and interdependent the world is due to the fast-global spread of the virus and the disruption to the world economy. The outbreak also triggered layoffs, loss of income, fear, and heightened uncertainty in all sectors of the global economy including education. These disruptions had a significant impact on trading partners and the global supply chain with an estimated loss of $9 trillion in 2020, according to the International Monetary Fund (Mou, 2020). Several countries reacted to control the spread of the virus by introducing lockdowns and travel restrictions. The pandemic also demonstrated the importance of international collaboration in the rapid production and distribution of personal protection equipment. It was followed by the rapid development, manufacture, and distribution of several vaccines in numerous countries around the globe. Scientists and engineers from various regions of the world worked together to make this possible.

Prior to globalization and the Internet era, most engineers were typically required to be technically competent and familiar with their local codes and standards since there were limited international collaborations. With the changing global market and ease of Internet connection, engineers must now be able to adapt culturally, be aware of global trends, and be prepared to innovate and take on new leadership roles. There are also ongoing challenges to address the uncertainties associated with the changing climatic conditions, ensure resilient infrastructures, and understand the associated costs and impact on local and global business practices. It has become apparent that most undergraduate engineering programs have been introducing students to professional practice in a limited context of the global economy, environment, and climate change. For future engineers to be successful in an increasingly global environment, they need to have an awareness of engineering needs and practices across various cultural, religious, ethical, economic, and political systems within the global

DOI: 10.1201/9781003280057-1

economy. As the social and cultural dimensions of today's globalization have led to an increased interconnection of people, ideas, and professional engineering practices, future engineers must develop sensitivity and understanding for other cultures in order to practice engineering within those settings.

Globalization processes have improved and continue to affect many areas of human life including education for professional practice in many disciplines. Understanding the effects of globalization on both teaching and learning is essential to educate professionals, who would become valuable and indispensable contributors to the integrated and interdependent global economy. Peters et al. (2020) reported that the COVID-19 pandemic heightening practices have already begun to emerge worldwide in higher education especially in the use of digital technologies. In the wake of the COVID-19 outbreak, universities around the world took measures to deal with the pandemic including temporary shutdown of campus operations and transition to online instruction. During the 2020–2021 academic year, most universities adopted hybrid models and suspended study abroad programs. This shift to online classes was plagued with multiple challenges such as bandwidth and Internet connectivity issues, hardware and software upgrades, faculty and staff expertise, as well as student engagement, motivation, and mental health. The pandemic offered educators throughout the world the opportunity to rethink not only new digital, online, and pedagogical possibilities but also to explore the basic purposes of education and how renewed vision of education might be harnessed to improve society. Utilizing the appropriate technology in higher education system would gradually blur the boundaries between countries and facilitate higher education to reach everyone or be accessed by everyone in the world. Therefore, college graduates must be prepared to meet the demands of globalization by embracing their critical roles in society as the world economy becomes more integrated and increasingly exposed to new challenges. With sound technical knowledge and a good understanding of cultural, environmental, business, and ethical forces within the global context, engineering graduates will be better prepared to work effectively in addressing the challenges within the changing global environment. Engineering students must understand the importance of leadership and creativity on professional practice and cultural, religion, political, legal, and economic systems in the global economy. Engineering professionals should strive to become responsible global citizens who are able to embrace and value the contributions of others and celebrate the cultural differences, and confidently move across cultures in the virtual or physical worlds in order to manage the complexities of living in a globalized world.

The National Academy of Engineering (NAE, 2004) in one of its publications *The Engineer of 2020: Visions of Engineering in the New Century* identified the importance of having future engineers gain knowledge and experience in topics such as the principles of leadership and management, ethical standards, cultural diversity, global/international impacts, and cost–benefit constraints. Preparing future engineers with technical and professional skills is critical to ensure the future development of the global professional engineering practice. Typically, engineers are not routinely prepared to lead groups of people and organizations through academic training or work assignments early in their career. The general concern is that leadership and management skills are not adequately addressed in traditional engineering undergraduate programs. Therefore, in response to feedback from the practicing engineering professionals, the American Society of Civil Engineers (ASCE) articulated new standards for civil engineering programs to help future engineers acquire the professional skills that are included in the ASCE's Body of Knowledge (BOK3, 2019). These professional components include proficiency in communications, critical thinking and problem-solving, public policy, sustainability, risk assessment, globalization, leadership, teamwork, professional attitude, life-long learning, and professional and ethical responsibilities. For more than two decades, ASCE has promoted the need for resilient civil engineers in the 21st century by calling on academic institutions to revise their approach to undergraduate engineering education. ASCE is advocating for industrial relevance, more practical exposure to real-life problems, mentored experience, development of professional skills, and the infusion of global awareness throughout the civil engineering curriculum. Most universities are acknowledging this trend and are working toward developing appropriate solutions. Part of the current academic approach includes offering international co-op or internship experience, partnership with industry, practical capstone projects, international exchange programs, and participation in global organizations such as Engineers Without Borders. In addition, students are encouraged to take more non-technical courses in humanities, social science, languages, business, and management.

Accreditation and assessment of engineering programs are vital to maintain the quality and the status of engineering graduates. It has been reported that there is a need for a systematic global model of engineering accreditation that can be used to assess global professional skills and attributes of engineering graduates. Patil and Codner (2007) presented a brief review of existing accreditation systems worldwide as well as proposed a global accreditation model for engineering education. Patil and Codner developed a three-dimensional competencies model for engineering graduate that includes assessment criteria of "Global Competencies" along with the "Hard and Soft Competencies" in the accreditation framework of engineering programs since graduates are expected to work within multicultural and multinational workplace environments. The Engineering Accreditation organization (Accreditation Board for Engineering and Technology, ABET) in the United States places emphasis on the successful preparation of undergraduates who are equipped with adequate skills needed for the practice of engineering in the 21st-century global marketplace. The ABET criteria expect the curricular content of engineering programs to include a general education component that is consistent with the Program Educational Objectives. To facilitate this, over two decades ago, ABET developed EC2000 criteria such as Student Outcomes "a-k" that evaluate attributes for entering the practice of engineering. Revised Student Outcomes (1 through 7) were adopted by ABET in 2018 intended to assure a systematic improvement of the quality of engineering education in a dynamic environment. In addition, ABET requires that civil engineering programs provide a framework of education that prepares graduates to enter the professional practice of engineering who are (i) able to participate in diverse multicultural workplaces; (ii) knowledgeable in topics relevant to their discipline, such as usability, constructability, manufacturability and sustainability; and (iii) cognizant of the global dimensions, risks, uncertainties, and other implications of their engineering solution.

Therefore, it is the responsibility of academic institutions to ensure that students have enough opportunities to develop and improve both technical and non-technical skills required for successful engineering practice throughout the world. The curricula of engineering programs should be strategically infused with perspectives that are more global to better prepare college graduates to meet the demands of globalization by embracing their critical roles in society.

This chapter highlights strategies that have been used or considered in educating globally minded and resilient engineers by infusing global perspectives, industry relevance, and leadership development into the civil engineering curriculum. The necessary global perspectives that will foster respect and understanding can be achieved by incorporating appropriate pedagogy and curricular developments within engineering programs. This approach that has been successfully used by the Civil Engineering Program at the United States Coast Guard Academy (USCGA) is discussed as a case study example. The focus is on how students develop as ethical leaders with an appreciation for global issues by infusing key aspects affecting the global engineering community into the curriculum and co-curricular activities.

1.2 DEVELOPING GLOBAL COMPETENCY AND RESILIENCE

With sound technical knowledge and understanding of cultural, environmental, business, and ethical forces within the global context, engineering graduates will be better prepared to work effectively in addressing the challenges within the changing global environment. Engineering students should demonstrate global competency; in particular, they should understand the importance of leadership and creativity on professional practice, cultural, political, and economic systems in the global economy. Colleges and universities have long sought to prepare their graduates to become responsible and informed global citizens; this increased awareness of the importance of global issues means that a college education is to help students develop an understanding of multiple interacting factors that span across social, economic, political, and environmental challenges. A wide range of perspectives and practices in higher education has emerged reflecting a considerable interest in empowering students with the knowledge in human conflict and cooperation, fostering cross-cultural understanding, and promoting the knowledge, attitudes, and skills relevant to living responsibly in a multicultural, interdependent world. Reimers (2009) argued that students require global skills and competencies, the attitudinal and ethical dispositions that make it possible to interact peacefully, respectfully, and productively across political, religious, and cultural borders. Similarly, Merryfield (2008) argued that students need to develop an acceptance of different cultures, a concern for the world, an understanding of interconnectedness, and the value of world citizenship. Rather than shifting emphasis into standardized knowledge of content and mastery of routine skills, many of the advanced educational systems are focusing on flexibility, creativity, and problem-solving through modern methods of teaching to develop deep learning, information literacy, and technical proficiency.

Parkinson et al. (2009) addressed the problem of defining global competence by exploring two research questions: (1) what does it mean to be globally competent? and (2) which dimensions of global competence are most important? Parkinson and his team explored the myriad of definitions for "global competence" and developed 13 dimensions that define what it means to be globally competent. The 13 dimensions of global competence for engineering graduates suggested by Parkinson et al. are as follows:

1. Appreciation of other cultures.
2. Ability to communicate appropriately across cultures.
3. Understanding of cultural differences relating to product design, manufacture, and use.
4. Experience in practicing engineering in a global context.
5. Effective resolution of ethical issues arising from cultural or national differences.
6. Proficient working in or directing a team of ethnic and cultural diversity.
7. Proficient in a second language at a conversational level.
8. Proficient in a second language at a professional (i.e. technical) level.
9. Informed on the history, government, and economic systems of other countries.
10. Understanding of the connectedness of the world and the working of the global economy.
11. Understanding the implications of cultural differences on how engineering tasks might be approached.
12. Exposure to international aspects of topics such as supply chain management, intellectual property, liability and risk, and business practices.
13. Appreciation of role as "Citizens of the world" – appreciate challenges facing humankind such as sustainability, environmental protection, poverty, security, and public health.

Klein-Gardner and Walker (2011) built on Parkinson et al.'s (2009) study by conducting a larger survey of industry and academia to clearly define what it means for engineering graduates to be globally competent. The survey results further validated the global competencies highlighted by Parkinson et al. (2009). Currently, industry requires engineers to adapt quickly and become resilient in the global environment. Resiliency can be defined as the ability to recover from or adapt to adverse conditions. In the same context, an engineer who is resilient (*the Resilient Engineer*) should demonstrate skills and attributes such as:

- Technical aptitude and commitment to life-long learning
- Adaptable leadership skills and ability to work in diverse and multidisciplinary environments
- Unwavering commitment to professional ethics across borders
- Capacity to handle change and adjust appropriately
- Cultural sensitivity and adaptability
- Ability to understand global perspectives of engineering practice
- Ability to recognize the influence of political environment and public policies

Therefore, a "global resilient engineer" can be defined as someone who has the personal qualities of adaptable teamwork and leadership, critical thinking,

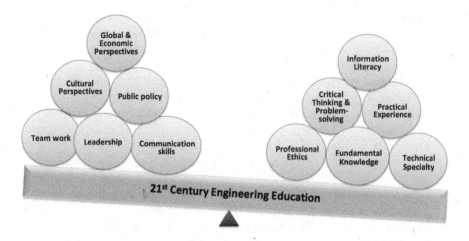

FIGURE 1.1   Attributes of a global and resilient engineer.

problem-solving, language and cultural awareness, effective communication, knowledge, and technical expertise to work effectively in diverse international settings and work environments. Combining these expected attributes of global competencies and resiliency then becomes a balancing act for engineering education as illustrated in Figure 1.1.

1.3   GLOBAL PERSPECTIVE IN THE CIVIL ENGINEERING PROGRAM AT USCGA

There are several principal ideas that should be included within the dimension of infusing global perspectives to meet the challenges of globalization in academe. Efforts should be made to address principles related to economic and political development, the environment, law, public policy, culture, gender and race equity, health, terrorism, peace and conflict resolution, rights, and responsibilities.

USCGA is a small undergraduate institution of approximately 1,000 cadets (students) with nine majors and 15%–20% of the cadet corps graduate with a Civil Engineering degree. The USCGA approach is to help students understand the nature of this interdependence and provide students with opportunities to study various interacting global systems that include economic, cultural, social, political, environmental, and technological aspects. Students are introduced to the strategies, skills, and dimensional issues that define global affairs. This approach has helped students understand how they are affected by the global systems and how they can develop strategies to live and work within those global systems. Students are encouraged to explore their roles in developing solutions to the issues and problems that exist within the global economy. To improve the preparedness of engineers entering the civil engineering practice and address global education, ASCE has adopted the attainment of a Body of Knowledge (BOK) for entry into the professional practice of civil engineering which includes: (1) foundational outcomes, (2) engineering fundamentals outcome, (3) technical outcomes, and (4) professional outcomes (ASCE BOK3,

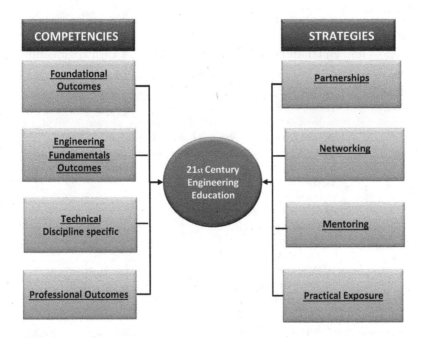

FIGURE 1.2   Proposed framework for engineering education.

2019). Technical dimension, as proposed in Figure 1.2, includes the foundational, engineering fundamentals, and technical outcomes described in the ASCE BOK. This includes competencies in mathematics, natural science, social science, humanities, technical specialty, critical thinking and problem-solving, and design experience. Although this model was developed for civil engineering, various components can be adopted by other engineering disciplines. Figure 1.2 is an illustration of the proposed framework that incorporates the key outcomes of ASCE-BOK3 that will promote global awareness in engineering education. Also included in Figure 1.2 are proposed strategies that may be helpful to foster and enhance global and industrial perspectives in engineering education.

A summary of how this framework was implemented by the Civil Engineering Program at USCGA is presented in Table 1.1. This summary includes examples of how the various competencies in Figure 1.2 were implemented within the curriculum.

The USCGA has adopted a holistic model to developing a resilient global engineer that is illustrated in Figure 1.3 and discussed in detail in the following sections. This model is dynamic and incorporates four key components, *Technical, Ethics, Cultural, and Leadership* that are complemented or supported by four professional attributes (information literacy, public policy awareness, global awareness, and service).

1.3.1   TECHNICAL COMPONENT

The Civil Engineering curriculum at USCGA includes a variety of required core courses in the humanities, science, engineering, mathematics, professional maritime

TABLE 1.1
Examples of Implementation Strategies Used in USCGA Civil Engineering

Component	Examples of Implementation
Foundational and engineering fundamentals	Competencies are achieved through technical engineering courses & core courses in math, science, humanities, business, law, maritime studies, management, engineering mechanics, engineering lab courses, critical thinking and problem-solving, and physical education.
Technical	Addressed through upper-level civil engineering courses (breadth and depth), semester design projects in various courses, and comprehensive capstone experience.
Professional	Addressed through particular emphasis in technical/core courses & co-curricular activities in areas such as: • Leadership and ethics • Communications • Teamwork • Community service • Professional training • Professional Engineering registration • Involvement in professional organizations • Information literacy • Global perspectives
Partnership & networking	Collaboration with several organizations & institutions such as: • Local universities and other military academies • Local non-profit/community programs • United States Coast Guard • Department of Homeland Security • Department of Defense • American Society of Civil Engineers • Society for American Military Engineers • Engineers Without Borders • Habitat for Humanities
Practical exposure	Emphasis is placed on infusing industrial and community relevance and Coast Guard mission readiness into the curriculum through practical class projects, field trips, guest speakers, capstone projects, conference, seminars and workshops, community service, summer internships, and membership in professional organizations.
Mentoring	Each student is assigned three mentors; academic faculty member, coach in athletic, and a military officer. Students are also required to mentor their juniors. There are also opportunities to interact with CG Chaplains and the local Sponsor Family Program that provides additional opportunities for individual growth.
Accreditation	The Civil Engineering Program is ABET accredited and continues to meet all ABET requirements in order to maintain this engineering accreditation. In addition, USCGA is also accredited by the regional higher education accreditation agency.

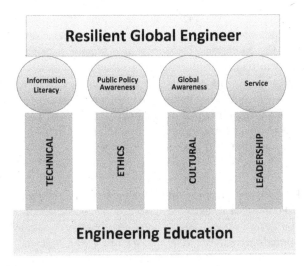

FIGURE 1.3 USCGA's global engineer development framework.

studies, organizational behavior, management, leadership, law, global awareness, and cultural perspectives. Students must successfully complete 67 semester credits of core courses, 55 semester credits of engineering courses, and at least 6 health and physical education credits. The Program provides students with a solid background in technical concepts and principles, problem-solving and critical thinking, understanding of professional responsibilities, leadership and ethical development, and exposure to other cultures and to political principles. Emphasis is placed on balancing theory and practice of engineering so that graduates are intellectually and professionally prepared to provide engineering services to the Coast Guard and the civil engineering profession. Professional skills are particularly reinforced in the engineering courses through laboratory reports, technical papers, oral presentations, design projects, field trips, interaction with practitioners and Coast Guard officers, community outreach activities, and professional membership. In addition to academics, students are also challenged through daily non-academic interactions with faculty and staff and through structured military and athletic training opportunities. There are also multiple opportunities each year for students to interact with local, federal, and international officials.

The development of technical competency is fostered by weaving experiential modes of teaching and learning throughout the undergraduate engineering courses. The teaching environment creates an atmosphere where students with a variety of learning styles can understand concepts and see the application of what they are learning. These techniques have been used to foster the development of technical competency as well as to infuse global and industrial awareness in the Civil Engineering Program. Examples of some of the pedagogical techniques are as follows:

- Combining classroom instruction with industrial exposure through field trips and summer internships. Each upper-level course is linked to one or more field trips. The objective of these trips is to reinforce topics discussed in class through exposure to practical applications. They provide familiarization with equipment, operation, production management, and problem-solving in real-life conditions. Field trips also provide excellent opportunities for students to interact with practicing professional and get a feel for real engineering work. This concept has been used at USCGA for more than a decade in which all civil engineering students participate in ten or more industrial field trips during their junior and senior years. There is also a program where students work/intern at Coast Guard Civil Engineering Units for 5 weeks between their junior and senior years.

- The design process is progressively integrated from freshman through senior year, and these design experiences have helped students perform at higher levels of cognitive learning. Project Based Learning (PBL) has been integrated into upper-level courses. An example of the use of PBL is through a common design project that is integrated into three senior-level courses (*Geotechnical Engineering*, *Reinforced Concrete Design*, and *Construction Project Management*). This is discussed in more detail in Chapter 2. Various project components are addressed in the appropriate courses so that students appreciate the connections between the various sub-disciplines in real-world civil engineering practice. Assessment data indicate that students are making progressive improvements in problem-solving abilities and are better prepared to complete their senior capstone design projects.

- Discussion of local and/or international projects as case studies and practical examples. Case studies provide opportunities for discussion of engineering principles and concepts as well as promote professional development in ethics and life-long learning. The use of case studies enhances students' understanding of engineering principles and improves their understanding of the problem-solving process. Several course instructors have integrated a "professional practice moment" in which students take turns presenting a contemporary engineering event during the first few minutes of each class.

- Several presentations by practitioners are planned throughout the program. Topics of discussion are linked to the course content and practical applications of the concepts. These presentations enable students to see the relevance of the concepts being taught and provide insight into the profession.

- During the COVID-19 pandemic, faculty revised their methods of instruction to facilitate remote learning as well as accessibility to course materials. This included the use of several platforms such as OneNote, MS Teams, Desire-to-Learn, and Camtasia. Faculty recorded short videos of lectures and labs that were accessible on multiple platforms. Exams were also revised into "take-home" and online formats. Students were given multiple opportunities to resubmit the required assignments and were encouraged to continue to be engaged in the learning process despite the stressful and difficult situation.

There is a strong collaboration between the civil engineering faculty and Coast Guard personnel at field units to identify real engineering projects for use as capstone design projects. This collaborative approach ensures that the Program fulfills its dual role of developing Coast Guard Officers and Civil Engineers with a thorough design experience that involves working on actual civil engineering projects with engineers in the field. Whenever possible, non-Coast Guard projects that are community based are included as capstone projects. Through community outreach with organizations such as Habitat for Humanity and local community programs, students have several opportunities for public service.

1.3.2 LEADERSHIP COMPONENT

Leadership development has become increasingly important in engineering education as leadership, communication, and teamwork skills are vital for success in professional engineering practice. To become effective civil engineers, students need to understand and develop skills or competencies in both leadership and management. Skipper and Bell (2006) recognized that leadership is a complex subject that is impacted by variables such as skill requirements at different stages in a career, the varying role assumed by leaders within different organizations, and the impact of technology. DeLisle (1999) defined engineering leadership as a direct function of three elements of interpersonal effectiveness: (1) people's awareness of themselves and other people; (2) their ability to make decisions, solve problems, motivate others, and balance the tasks and relationships in an organization; and (3) the commitment to make hard decisions and face the risk of "doing the right thing." He suggested several traits engineers must develop to become effective leaders: (1) seek feedback and information about their interpersonal effectiveness, (2) develop understanding of how people are motivated and how they grow and develop in the organization, (3) develop and sustain conceptual flexibility and be comfortable with change and ambiguity, and (4) hone technical problem-solving skills but develop an equally competent set of interpersonal skills related to communications.

Leaders must practice mentoring, must be willing to work with others as a team, and assist the nation in finding new approaches for rebuilding the infrastructure for enhanced quality of life and global competitiveness. Perry et al. (2017) proposed a theoretical framework system and actionable recommendations for an educational model designed to transform engineers into leaders during their undergraduate and/ or graduate education experience. It is well known that skill sets required for engineering jobs and leadership roles are often distinct; engineers learn technical principles required for their specific engineering discipline, whereas effective leadership requires strategy, communication, persuasion, motivation, and a myriad of people skills. The proposed principles are about shifting mindset and designing an environment with opportunities to capitalize on that mindset shift, meanwhile providing everyday, real-world leadership development opportunities for more engineers, even when they would not self-select into a leadership development program. Perry et al. (2017) proposed an Organizational Innovation Model for Education that contains three strategic pillars:

- *Channeled Curiosity* – This covers four principles: Lead strategically with vision, adopt a platform mentality, adopt a synthesis mentality, and persist.

- *Boundary Breaking Collaboration* – This involves leading through persuasion and trust, creating interdependence, and building bridges across boundaries.
- *Orchestrated Commercialization* – This involves coordinating networking, elevating role models, and revisiting incentives.

Helping students develop such skills is not an easy task. A good start is to have academic institutions make developing professional attributes as one of their missions by providing opportunities to develop leadership skills through coursework and extracurricular activities. Through coursework, students can be exposed to established principles of leadership and these can then be applied by participating in extracurricular activities. Leadership development at USCGA is based on the USCG Commandant Instruction COMDTINST M5351.3 (2006) which is divided into four categories: *leading self, leading others, leading performance and change*, and *leading the Coast Guard*. These four categories consist of several competencies or skills such as team building, communications, accountability, conflict management, and technical proficiency that students develop over 4 years at USCGA. The competencies that students develop through this process are summarized in Table 1.2. During their freshman year, the focus is predominately on the competencies detailed in the leading-self category. Freshman year is one of the most important stages of a student's academic life. This is when much of self-discovery, understanding of the academic system, and an early development of technical proficiency in the areas of study and learning take place. With each progressive year, the students shift their focus to the next set of competencies, while still working on honing their skills in the previous categories. The academic component of leadership development at USCGA is supported by a strong "Core Curriculum" of science, math, engineering, professional

TABLE 1.2

USCGA's Leadership Development Framework

Leading Self	Leading Others	Leading Performance and Change	Leading the Coast Guard
Accountability and responsibility	Effective communications	Conflict management	Financial management
Aligning values	Team building	Customer focus	Technology management
Followership	Influencing others	Decision-making and problem-solving	Human resource management
Health and well-being	Mentoring	Management and process improvement	External awareness
Self-awareness and learning	Respect for others and diversity management	Vision development and implementation	Political savvy
Personal conduct	Taking care of people	Creativity and innovation	Partnering
Technical proficiency	Technical proficiency	Technical proficiency	Entrepreneurship, stewardship, strategic thinking

studies and humanities courses. In addition to academics, students are challenged non-academically (militarily, athletically, and socially) through daily interactions with each other, USCGA faculty, and staff and through structured military and athletic training opportunities. This framework is discussed in more detail in Chapter 3.

1.3.3 ETHICS COMPONENT

Ethics education in civil engineering undergraduate programs has been ongoing for years per ASCE Code of Ethics (2019). However, in the last two decades, there has been a growing emphasis on the importance of ethics education in engineering. Professional ethics education will not by itself reduce unethical practices, but awareness of proper conduct and the empowerment of individuals are critical outcomes of a professional ethics educational program. This contributes significantly to the ongoing preservation of ethical behavior in any profession. A comprehensive professional ethics education will support individuals to become critically aware and scrutinize practices around them rather than becoming enculturated with unethical existing norms, practices, and values. Tarhini et al. (2015) suggested that an important factor in developing students' ethical reasoning ability is exposing them to curricular experience that require deep cognitive processing about ethical issues at the appropriate level of Bloom's Taxonomy. Carpenter et al. (2014) recommended promoting extracurricular and co-curricular ethical involvement through service programs and professional organizations. Institutions should establish and communicate clear behavioral expectations across administration, faculty, staff, and students such that the culture of the institutions promotes ethical development.

The USCGA's mission: "To educate, train and develop leaders of character who are *ethically*, intellectually, and professionally prepared to serve their country and humanity." The Academy articulates core values that include: *Honor, Respect, and Devotion to Duty*; these values are the framework for the professional conduct and ethical responsibilities that the Coast Guard expects. USCGA expects students to conduct themselves in accordance with an Honor Concept which requires that "Students neither lie, cheat, steal, nor attempt to deceive." Everyone must integrate this concept into their way of life and foundation for interactions with all persons. Breaches of the Honor Concept are serious offenses that normally result in disenrollment from the Academy. Details on how ethics is addressed in the Civil Engineering Program at USCGA are presented in Chapter 5.

1.3.4 CULTURAL/DIVERSITY COMPONENT

Developing global awareness involves learning about problems and issues that cut across national boundaries and about the interconnectedness of systems – social, cultural, economic, political, and technological. The social and cultural dimensions of today's challenges have led to an increased interconnection with people, goods, and ideas. Only with an understanding and appreciation of practices across cultures will engineering graduates be prepared to serve in a global market while fostering respect and tolerance for differences in culture, race, and religion. The effectiveness of engineers in global practice depends on how successfully cultural differences are addressed and understood. Undergraduate engineering students should be introduced to the strategies, skills, and dimensional issues that define global affairs. Three main aspects must be addressed by engineering programs:

1. The engineering community must be knowledgeable about the opportunities and the benefits that international markets provide.
2. Universities and industry must collaborate to improve the engineering curricular programs to fully take advantage of these opportunities.
3. Governments are encouraged to make policies that support academic initiatives which would prepare future engineers to be culturally competent and successfully function in the global marketplace.

The student body at USCGA is made up of students from almost every state in the United States of America (USA) as well as international students. There is a rich culture of teamwork, comradery, and mutual respect. Every effort is made to ensure that students are developed in an inclusive and supportive environment. This environment is cultivated and fostered through diversity councils and activities supported by the Office of Inclusion and Diversity. Some of the activities that promote cultural understanding and diversity across the Academy include:

- *Women's Leadership Council* – This council promotes female health and wellness as well as fosters a professional female leadership network. The focus is on issues that affect the female workforce with an emphasis on highlighting the current leadership roles of women in the Coast Guard. The council activities best prepare members for challenges and experiences they will face in the Coast Guard while building leaders of character.
- *International Council* – This organization is run by international students with supervision from two faculty advisors and the Office of Inclusion and Diversity. Membership is open to the entire student body. The International Council engages students in activities that promote global cross-cultural exchange, and dialogue by expanding their view of the world and their roles as future leaders. This is accomplished through several activities that prepare students for effective participation in the dynamic global society, broaden their awareness of other international cultures, develop international networks, and establish relationships that could be leveraged in the future. Some of the Council's annual activities include community service, visits to the United Nations Headquarters, the Naval and Army Academies, speed mentoring program, and international culture nights.
- *Genesis Council* – This council provides students with opportunities to learn about and experience, first-hand, the history, cultural nuances, and issues that characterize the African American and Black diaspora. It provides a support network that facilitates a familial atmosphere for interested students to fellowship with each other. The Council promotes activities that bring students from diverse backgrounds together for positive discourse on contemporary racial issues, intellectual stimulation, and personal growth.

- *Compañeros Council* – This council provides a channel through which students can celebrate Hispanic heritage and promote awareness of the Hispanic culture and norms. The Council strives to improve community relations by engaging in service and local cultural events including promoting Hispanic cuisine, dances, films, and the Spanish language.
- *Asian-Pacific American Council* – This council provides students numerous opportunities to become better-informed future leaders who are more prepared for the growing diversity in the Coast Guard as well as the American workplace. This is accomplished by sponsoring cultural events, community service, guest speakers, social activities, and networking with other schools that promote a better understanding of the Asian and Pacific Islander cultures.
- *Eclipse Event* – During the spring semester, the Office of Inclusion and Diversity together with faculty advisors and the leadership of eight student councils plan a major event full of various activities that focus on generating a climate of inclusion, promoting tolerance and cultural understanding. Several Coast Guard officers, alumni, and local and national experts on diversity are invited to mentor and inspire students to become global citizens. Eclipse week supports USCGA's Strategic Plan vision of "Modeling a Community of Inclusion," improving retention and strengthening recruitment and officer accession.
- *Travel Abroad Opportunities* – Although limited, some students have opportunities to travel abroad as part of their summer assignments, mini study abroad semester, and capstone projects. There have also been overseas travels to participate in international projects that involve the interest of the United States and the Coast Guard. The most traveled areas include the Arctic region, European Union, and Africa. Students also have opportunities to participate in national and international conferences as well as regional meetings.

Another opportunity for student exposure to key cultural challenges in a global context is provided by the study of the Holocaust. The study of the Holocaust opens up a platform for discussions about the importance of leadership, silent followership, and respect of diversity in organizations. Students are expected to recognize the ethical implications of the Holocaust and understand that atrocities occur because people and governments make decisions that perpetuate discrimination and persecution. Group discussions and study of the Holocaust enable students to think about political responsibilities and explore the functioning of governmental structures. An analysis of the mechanisms that led to the Holocaust helps in recognizing the importance of accepting and appreciating diversity rather than seeing it as a cause for discord.

1.4 CLOSING THOUGHTS

As the current rapid social and economic changes caused by globalization continue, it is vital that institutions of higher education consider their contribution to the global society from a broad and long-term perspective. To meet the challenges associated with these changes, academic institutions must focus on the development of a skilled and knowledgeable workforce with the appropriate global perspective that will be resilient in such an environment. It is critical that students develop the skills and abilities that will promote cultural sensitivity, respectful international relations, ethical and moral values, teamwork, service, and adaptive leadership. USCGA has created an environment with institutional goals for student learning, growth, and development both inside and outside the classroom to ensure that graduates are prepared to meet the many challenges they will face throughout their professional careers. The Civil Engineering Program, in particular, has integrated several aspects of globalization into the curriculum. Emphasis is placed on having students understand the multi-dimensional changing world around them. The multi-dimensional approach is based on a holistic model with four key components or dimensions (technical, cultural, ethics, and leadership) that are complemented by professional attributes such as global awareness, service, public policy awareness, and information literacy. The jury is still out on the key competencies of the engineer with the appropriate level of global awareness. However, engineers should appreciate the challenges facing the world such as sustainability, environmental protection, climate change, resiliency, poverty, security, and public health. Some good lessons have been learned during the COVID-19 pandemic, and universities may have to gear up to provide, in addition to regular education (in-person), online education lectures that can be accessed globally.

REFERENCES

ABET. (2018). *Criteria for Accrediting Engineering Programs.* www.abet.org. Board of Directors, Baltimore, MD.

American Society of Civil Engineers. (2019a). *Civil Engineering Body of Knowledge Preparing the Future Civil Engineer,* 3rd Edition, ASCE Press, Reston, VA.

American Society of Civil Engineers. (2019b). *American Society of Civil Engineers Code of Ethics.* http://www.asce.org/code_of_ethics/. Accessed March, 2019.

Carpenter, D., Harding, T., Sutkus, J. and Finelli, C. (2014). "Assessing the ethical development of civil engineering undergraduates in support of the ASCE body of knowledge." *ASCE Journal of Professional Issues in Engineering Education & Practice,* Vol. 140, No. 4, on-line A4014001.

Commandant Instruction COMDTINST M5351.3. (2006). *Leadership Development Framework.* United States Coast Guard, Washington, DC.

DeLisle, P.A. (1999). "Engineering leadership, the balanced engineer: entering a new Millennium." *Proceedings of IEEE-USA Professional Development Conference,* Dallas, TX.

Klein-Gardner, S. and Walker, A., (2011). "Defining global competence for engineering students." *ASEE Annual Conference & Exposition,* Vancouver, B.C., Canada, June 26–29, 2011.

Merryfield, M. (2008). "The challenge of globalization: preparing teachers for a global age. Teacher education & practice." *The Journal of the Texas Association of Colleges for Teacher Education,* Vol. 24, No. 4, pp. 435–437.

Mou, J.J. (2020). "Research on the impact of COVID19 on global economy." *IOP Publishing Conference Series: Earth and Environmental Science,* 546, AEECE. Doi: 10.1088/1755-1315/546/3/032043.

National Academy of Engineering. (2004). *The Engineer of 2020: Visions of Engineering in the New Century*. National Academy Press, Washington, D.C.

Parkinson, A., Harb, J. and Magleby, S. (2009). "Developing global competence in engineers: What does it mean? What is most important?" *ASEE Annual Conference & Exposition*, Austin, TX, June 14–17, 2009.

Patil, A. and Codner, G. (2007). "Accreditation of engineering education: review, observations and proposal for global accreditation." *European Journal of Engineering Education*, Taylor & Francis Group, Vol. 32, No. 6, pp. 639–651.

Perry, S., Hunter, E., Currall, S. and Frauenheim, E. (2017). "Developing engineering leaders: an organized innovation approach to engineering education." *Engineering Management Journal*, Taylor & Francis Group, Vol. 29, No. 2, pp. 99–107.

Peters, M.A. et al. (2020, June 25). "Reimagining the new pedagogical possibilities for universities post COVID-19." *Journal of Educational Philosophy and Theory*, Taylor & Francis, Doi: 10.1080/00131857.2020.1777655.

Reimers, F.M. (2009). "Leading for global competency." *Teaching for the 21st Century*, Vol. 67, No. 1, pp. 1–7.

Skipper, C.O. and Bell L.C. (2006, April). "Influences impacting leadership development." *ASCE Journal of Management in Engineering*, Vol. 22, No. 2, pp. 68–74.

Tarhini, K., Jackson, H., Zelmanowitz, S. and Zapalska, A. (2015). Developing ethical engineers at the U. S. coast guard academy: integrating courses, co-curriculum activities, and a global perspective. *ASEE-Northeast Section Conference*, Northeastern University, Boston, MA, 30 April – 02 May, 2015.

2 Critical and Design Thinking across the Civil Engineering Curriculum

2.1 INTRODUCTION

The engineering education landscape continues to change due to technological advancement, industry expectations, and global trends. There have been several calls from the engineering industry to reform undergraduate education so that graduates are better prepared to address the challenges posed by this changing landscape. Educational institutions are adapting by revising their programs so that tomorrow's engineering professionals have the technical knowledge and skills necessary to meet industry expectations. Fostering the development of critical and design thinking skills has become one of the key goals of college education where a variety of pedagogic techniques are being used to develop these skills in undergraduate students. To be successful, engineering professionals must be technically competent, culturally aware, and be adept critical thinkers and problem solvers.

Generally, engineering undergraduate programs are required to meet accreditation criteria that ensure students develop an appropriate level of critical- and design-thinking skills. In the United States, these skills are encompassed within accreditation criteria such as Accreditation Board for Engineering and Technology (ABET) Student Outcomes. As such, design and critical thinking in engineering education tend to mostly occur in a focused context directed toward fulfilling one or more ABET Student Outcomes. Critical thinking skills allow individuals to approach specific problems, questions, and issues with clarity, orderliness, diligence, persistence, and precision. Engineering educators have introduced critical thinking and design in the context of problem-solving, ethical decision-making, open-ended design, and assessing the social impacts of technology. Even though engineering educators place emphasis on critical and design thinking, traditional teaching pedagogies do not consistently facilitate it. Lectures, multiple choice, and short answer problems typically used in traditional teaching do not necessarily give students enough opportunities to foster the development of various solutions or reflections. The ability for students to think critically fosters their intellectual development to: appropriately use information and knowledge, recognize and define problems, ask relevant questions, search a variety of sources for relevant information, identify potential solutions, select the best solution, and verify results that meet problem definition and constraints.

This chapter introduces teaching strategies used to infuse critical and design thinking development in engineering courses along with assessment tools to evaluate

DOI: 10.1201/9781003280057-2

students' progressive development across the curriculum. The assessment and skill development are based on the six cognitive levels of Bloom's Taxonomy. The strategy also illustrates how undergraduate engineering students can be helped to understand the importance and principles of design thinking through the progressive implementation of problem-solving and project-based teaching and learning across the curriculum.

2.2   BACKGROUND ON CRITICAL THINKING CONCEPTS

The mainstream approach to critical thinking is perhaps best captured in the Delphi Report prepared by Facione (1998) that was developed as a consensus document by a group of leading academics. The consensus statement developed by the group regarding critical thinking and the ideal critical thinker is

> We understand critical thinking (CT) to be purposeful, self-regulatory judgment which results in interpretation, analysis, evaluation, and inference, as well as explanation of the evidential, conceptual, methodological, criteriological, or contextual considerations upon which that judgment is based. CT is essential as a tool of inquiry. As such, CT is a liberating force in education and a powerful resource in one's personal and civic life. While not synonymous with good thinking, CT is a pervasive and self-rectifying human phenomenon. The ideal critical thinker is habitually inquisitive, well-informed, trustful of reason, open-minded, flexible, fair-minded in evaluation, honest in facing personal biases, prudent in making judgments, willing to reconsider, clear about issues, orderly in complex matters, diligent in seeking relevant information, reasonable in the selection of criteria, focused in inquiry, and persistent in seeking results which are as precise as the subject and the circumstances of inquiry permit. Thus, educating good critical thinkers means working toward this ideal. It combines developing CT skills with nurturing those dispositions which consistently yield useful insights, and which are the basis of a rational and democratic society.

The term "critical thinking" is familiar to most engineering educators, but it is difficult to agree on a definition. Scriven and Paul (1987) defined critical thinking as:

> the intellectually disciplined process of actively and skilfully conceptualizing, applying, analyzing, synthesizing, and/or evaluating information gathered from, or generated by observation, experience, reflection, reasoning, or communication, as a guide to belief and action. In its exemplary form, it is based on universal intellectual values that transcend subject matter divisions: clarity, accuracy, precision, consistency, relevance, sound evidence, good reasons, depth, breadth, and fairness.

From this definition, the three key elements of critical thinking are reason, reflection, and judgment. The combination of reflection and reason leads to the final elements, which are belief in the validity of a premise, solution to a problem, and an appropriate action. A good critical thinker develops conclusions by deducing or inferring answers to questions and then reflecting on the quality of the reasoning; the result is conviction, and in many cases action, based on those conclusions. Critical thinking is the combined process of analysis, synthesis, and evaluation. Critical thinking is interwoven with several modes of thinking such as engineering thinking, scientific thinking, mathematical thinking, and philosophical thinking. General critical

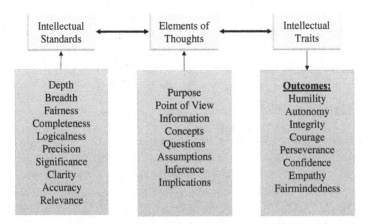

FIGURE 2.1   The critical thinking framework. (Adapted after Paul et al., 2013.)

thinking can be thought of as a scientific method that involves problem definition, the research for solutions, evaluation of solutions, and iteration when appropriate. This is similar to the engineering critical thinking process (which is usually referred to as the "Design Process") that also starts with a problem definition, followed by information gathering, analysis, evaluation, and production.

Paul et al. (2013) developed a critical thinking guidebook tailored to engineering educators and promoted the framework of engineering reasoning as shown in Figure 2.1. This framework depicts critical thinking by applying *Universal Intellectual Standards* to the evaluation of typical *Elements of Thought*, with the goal of developing certain *Essential Intellectual Traits* in the thinker. The framework allows for the analysis and evaluation of thought, but more importantly, it provides a common vocabulary for engineering educators. Essentially, the Universal Intellectual Standards are the criteria used to evaluate the quality of the critical thinking. According to this framework, applying the standards to the elements is what transforms general or everyday thinking to critical thinking. The overall goal is the development of the Essential Intellectual Traits that are characteristic of a maturing critical thinker. The focus of the framework is the eight Elements of Thought that clarify the building blocks of thinking; these building blocks can be used by anyone who examines, analyzes, and reflects on intellectual work. These elements are embodied in eight categories of inquiry that can be summarized: purpose, problem statement, assumptions and constraints, perspectives, evidence, and associated consequences.

Ahern et al. (2012) investigated the development of critical thinking in the university curriculum in Ireland. The research included semi-structured interviews of students from various disciplines. They found that critical thinking is not a static attribute that all students should aspire to as the ultimate destination in their education. Instead, critical thinking was found to be a dynamic concept that requires educators to guide students through engagement with context-bound knowledge and the empirical, on the one hand, and knowledge that is abstract and reflective, on the other. Critical thinking was found to be the movement back and forth between those two states of abstract and reflective. Ahern et al. defined critical thinking as

> a movement from the concrete, from the factual to the abstract and back again – an ability not only to use knowledge and facts to create ideas, concepts and solve problems, but also to use these developed concepts, theories, and ideas in the real world.

Therefore, Ahern et al. concluded that engineering educators must be more explicit and intentional in addressing critical thinking in the curriculum and give students clearer guidelines on how to become critical thinkers.

Adair and Jaeger (2016) outlined how critical thinking was introduced in an undergraduate engineering program and how it can be nurtured through practical experience with appropriate guidance and reinforcement. They demonstrated how critical thinking components relate to the principles of problem-solving procedure in a Fluid Mechanics course. They developed a method of assessment during the various phases of the project involving different levels of critical thinking. The methodology they used included the following steps:

- *Motivation* – I can do this/I want to do it.
- *Definition of the Problem* – Define problem/sketch/determine appropriate information or constraints and define criteria for judging answer.
- *Explore the Problem* – What is the real objective? What are the issues? What would be reasonable assumptions? Give an approximate answer.
- *Plan the Solution* – Develop overall plan, develop any sub-problems, select appropriate principles/approach, and determine any research that needs to be done.
- *Implement the Plan* – Apply the solution to solve the problem
- *Check the Solution* – Check the accuracy of calculations and check units.
- *Evaluate and Reflect* – Is the answer reasonable? Does it make sense? Were the assumptions good? How does the solution compare with an approximate answer? When appropriate, is the solution ethically sound?

Adair and Jaeger (2016) contrasted between "exercise-solving" classroom instruction and classroom instruction that focuses on developing students' critical thinking skills. A summary of their comparison between the two approaches is shown in Table 2.1.

Gude and Truax (2015) argued that critical thinking is exercised whenever a decision must be made on a problem that has more than one solution. Therefore, critical thinking requires reliable information and evidence to make an appropriate scientific decision. The critical thinking process was found to be a function of the individual's experience, technical expertise, basic intuition, and engineering reasoning. They defined the core components of critical thinking (as applied in engineering) as the "interpretation of an engineering problem that involves analysis, inference, explanation, evaluation, and self-regulation." Description of these critical thinking components is presented in Table 2.2. Gude and Truax developed and incorporated three methods to help students develop critical thinking skills in environmental engineering: (1) *problem-based learning* (solving single and open-ended solution problems);

TABLE 2.1
Critical Thinking Components Versus the Traditional Exercise Approach

Critical Thinking Delivery	Exercise Solving Delivery
Emphasizes the best answer to an unknown	Emphasizes only one right answer
Include ambiguity within given information	Well-defined situation with all information explicitly given
New situations given	Building on similar situations previously encountered
No clues as to what knowledge or skill needed	Hints and prescribed assumptions often given
There could be more than one valid approach	Typically one approach gives a correct answer
The method of solving is not clear	Build on familiar solutions
Knowledge necessary to solve problem can come from several subjects	Problems tend to come from one subject only and even only one topic
Require good communication skills	Communication skills not often required as equations/ graphs are all that is necessary

Source: Adapted after Adair and Jaeger (2016).

(2) *collaborative learning* (team- and project-based learning); and (3) *inquiry-based learning* (improves student learning of the subject matter through inquiry, discovery, evaluation, and problem-solving based activities). They concluded that:

- Problem-based learning method promoted students' critical thinking ability.
- Inquiry-based learning improved student learning of the subject matter through inquiry, discovery, evaluation, and problem-solving-based activities.
- Supplemental instruction gave students opportunities to apply skills used by practicing engineers and exercise critical (reflective) thinking in solving engineering design problems. This enabled students to identify confusion and misunderstanding, formulate the right question, apply previous knowledge and experience, and identify appropriate sources of information.
- The actual method that students apply to learn critical thinking is a mix of approaches that follow a progressive pattern. Progression from individual learning to small-group learning and classroom-level learning provides a good opportunity for discovering various aspects of the topic under discussion.

2.3 CRITICAL/DESIGN THINKING AND BLOOM'S TAXONOMY

Benjamin Bloom (1956) established a framework categorizing educational goals and objectives into a hierarchical structure representing different forms and levels of cognitive learning. The framework of Bloom's Taxonomy was published in *Taxonomy of Educational Objectives* (1956) in which the *Cognitive Domain* was defined as a knowledge-based domain with six levels of intellectual or thinking skills. The *Bloom's Taxonomy of Learning* classifies learning according to increasing levels of difficulty and complexity. In this hierarchical framework, each stage of learning is a precondition for the next stage, and, therefore, mastery of a given stage of learning

TABLE 2.2
Description of the Critical Thinking Components

Component	Primary Skills	Secondary Skills
Interpretation of an engineering design problem	Comprehend and express the meaning or significance of a wide variety of experiences, situations, data, events, judgments, conventions, beliefs, rules, procedures, or criteria.	Categorization, decoding significance, and clarifying meaning
Analysis	Identify the intended and actual inferential relationships among statements, questions, concepts, descriptions, or other forms of representation intended to express belief, judgment, experiences, reasons, information, or opinions.	Examining ideas, detecting arguments, and analyzing arguments
Evaluation	Assess the credibility of statements or other representations which are accounts or descriptions of a person's perception, experience, situation, judgment, belief, or opinion; and to assess the logical strength of the actual or intended inferential relationships among statements, descriptions, questions, or other forms of representation.	Judgment, assessment of the problem
Inference	Identify and secure elements needed to draw reasonable conclusions; to form conjectures and hypotheses; to consider relevant information; and to deduce the consequences flowing from data, statements, principles, evidence, judgments, beliefs, opinions, concepts, descriptions, questions, or other forms of representation.	Querying evidence, conjecturing alternatives, and drawing conclusions
Explanation	Able to present in a cogent and coherent way the results of one's reasoning. Provide a comprehensive view at the big picture: both to state and to justify that reasoning in terms of the evidential, conceptual, methodological, criteria-based, and contextual considerations upon which one's results were based; and to present one's reasoning in the form of cogent arguments.	Describing methods and results, justifying procedures, and presenting full and well-reasoned arguments in the context of seeking the best understanding possible
Self-regulation	Self-consciously monitor one's cognitive activities, the elements used in those activities, and the results deduced, particularly by applying skills in analysis, and evaluation to one's own inferential judgments with a view toward questioning, confirming, validating, or correcting either one's reasoning or one's results.	Self-examination and self-correction

Source: Adapted from Gude and Truax (2015).

requires mastery of the previous stage. Consequently, the taxonomy logically leads to classifications of lower and higher order levels of learning that is today considered a standard framework for learning, teaching, and assessment. This framework has been widely used by teachers, academic professors, and administrators. The six increasing levels of learning of the cognitive domain defined by Bloom in 1956 are as follows:

- *Knowledge* – remembering previously learned information; this is the first (lowest) level in Bloom's cognitive domain that involves identifying and defining the repetition of principles and concepts learned.
- *Comprehension* – grasping the meaning of information; this is the second level in the cognitive domain and involves a step beyond just memorization expressed in the first level. This level involves describing, explaining, summarizing, and understanding of the relevance of principles and concepts learned.
- *Application* – applying knowledge to actual solutions; this is the third level of cognitive domain and involves utilizing principles and concepts to address a problem or situation.
- *Analysis* – breaking down objects or ideas into simpler parts and seeing how the parts relate and are organized; this is the fourth level of the cognitive domain. It involves detailed scrutiny of principles and concepts, making the relevant and appropriate connections between several components and how they fit together to categorize a problem or situation.
- *Synthesis* – rearranging component ideas into a new whole; this is the fifth level of the cognitive domain. It involves formulating and integrating principles and concepts into well-organized processes to design as well as explain their appropriateness to address a given problem or situation.
- *Evaluation* – making judgments based on internal evidence or external criteria; this is the sixth (highest) level of the cognitive domain that involves assessing and investigating a situation or alternative solution(s) making a judgment on validity of solutions based on the applied principles and concepts.

Anderson and Krathwohl (2001) revised the previous order of the 1956 Bloom model by replacing the synthesis and evaluation with evaluating and creating as the highest order of thinking skills as presented in Figure 2.2. Traditional teaching methods such as lecturing usually engage students at lower levels of Bloom's taxonomy because the focus is mostly on downloading information without really fostering critical thinking. Restructuring the educational process to take students through higher levels of learning has been identified as crucial to cultivating life-long learning.

In engineering education, additional development in critical thinking is fostered through the design process. Miller and Linder (2015) defined Design Thinking as "an engaging process and methodical framework for approaching complex, multidisciplinary problems in ways that consistently result in solutions that are successful and often creative in unpredictable ways." Creating frameworks or perspective for design thinking takes time, patience, and the intentional design of classroom exercises and assignments that encourages students to practice design thinking sequentially through the six levels of Bloom's Taxonomy. With higher levels of critical and design thinking, students will significantly improve their ability to identify and understand

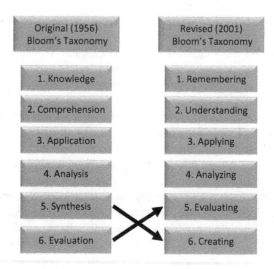

FIGURE 2.2　Original and revised bloom taxonomy.

a problem, determine key parameters, solve complex engineering problems, develop alternatives, and produce higher quality solutions. The design process challenges students to synthesize, analyze, evaluate, and apply the professional skills, knowledge, and tools they have acquired. While the design thinking and critical thinking skill sets are highly valued in engineering graduates, they are also the most difficult to learn and teach. Therefore, introducing design thinking and critical thinking in the freshman year and reinforcing it throughout the 4 years of undergraduate engineering education is beneficial for both students and faculty. The more frequently students are exposed to and given opportunities to apply the various components of the design process, the more critical thinking experience they gain before graduation.

Several research publications indicate that any learning activity that involves evaluating, analyzing, critiquing information as well as forming a hypothesis, collecting data, analyzing the data, and then making conclusions is one of the best techniques for critical thinking development (Ahern et al., 2019; Cotter & Tally, 2009; Fitzgerald, 2000; Elder & Paul, 2010). Instruction that advances design and critical thinking should be based on questioning techniques where students are able to evaluate information to solve problems and make decisions instead of repeating and memorizing information that they were expected to master (Johnson et al., 1998).

2.4　CRITICAL THINKING WITHIN ABET STUDENT OUTCOMES

Anyone thoroughly versed in the Paul-Elder model (2009) would argue that critical thinking is the foundation for all the ABET student outcomes accreditation requirements that engineering programs must demonstrate that their students achieve. Although the term "critical thinking" is not used in any outcome, they all require "the intellectually disciplined process of actively and skillfully conceptualizing, applying, analyzing, synthesizing, and/or evaluating information gathered from, or generated by, observation, experience, reflection, reasoning, or communication, as a guide to belief or action," i.e. critical thinking as per Scriven and Paul (1987). The following ABET Criterion 3 Student Outcomes in particular can be directly linked to critical and design thinking in civil engineering (ABET, 2018):

- ABET 3-1 "Ability to identify, formulate, and solve complex engineering problems by applying principles of engineering, science and mathematics."
- ABET 3-2 "Ability to apply engineering design to produce solutions that meet specified needs with consideration of public health, safety and welfare, as well as global, cultural, social, environmental, and economic factors."

The engineering community accepts that critical thinking is foundational to life-long learning and the assessment of critical thinking will complement the assessment of ABET outcomes for engineering programs. The American Society of Civil Engineers Body of Knowledge (ASCE-BOK3) developed a set of outcomes including development of critical thinking and problem-solving skills that are expected to be fulfilled through a combination of undergraduate education and mentored experience as illustrated in Table 2.3. The Civil Engineering Body of Knowledge comprises 21 outcomes in four categories that incorporate levels of achievement required for entry

---

TABLE 2.3
ASCE-BOK3 Foundation for Critical Thinking and Problem-Solving

Cognitive Domain Level of Achievement	Demonstrated Ability	Typical Pathway
Remember (remember previously learned material)	**Identify** and **define** a complex problem, question, or issue relevant to civil engineering.	Undergraduate education
Comprehend (grasp the meaning of learned material)	**Explain** the scope and context of a complex problem, question, or issue relevant to civil engineering.	Undergraduate education
Apply (use learned material in new and concrete situations)	**Formulate** a possible solution to a complex problem, question, or issue relevant to civil engineering.	Undergraduate education
Analyze (break down learned material into its components parts so that its organizational structure may be understood)	**Analyze** a possible solution to a complex problem, question, or issue relevant to civil engineering.	Mentored experience
Synthesize (put learned material together to form a new whole)	**Develop** a set of appropriate solutions to a complex problem, question, or issue relevant to civil engineering.	Mentored experience
Evaluate (judge the value of learned material for a given purpose)	**Assess** a set of solutions to determine the most appropriate solution to a complex problem, question, or issue relevant to civil engineering.	Mentored Experience

Source: ASCE-BOK3- 2019, reproduced with permission from ASCE.

into the civil engineering at the professional level (ASCE-BOK3, 2019). The cognitive domains shown in Table 2.3 are based on Bloom's 1956 taxonomy; the demonstrated abilities correlate well with ABET outcomes and the Paul-Elder's model shown in Figure 2.1.

2.5   CASE STUDY: EXAMPLES OF IMPLEMENTATION AT USCGA

Critical thinking development process within the Civil Engineering Program at the United States Coast Guard Academy (USCGA) is based on Bloom's Taxonomy shown in Figure 2.2. Critical thinking skills are developed sequentially and progressively under the guidance of each instructor in selected courses in the Civil Engineering Program. Students at various stages in their academic development are guided through the six levels of Bloom's cognitive learning from "Knowledge" through "Evaluation or Creating" levels of critical thinking skills. The Civil Engineering faculty are constantly challenging students to use their critical and design thinking skills progressively throughout the curriculum to achieve the highest cognitive level in Bloom's Taxonomy. The faculty uses performance indicators to assess design competencies that are mapped to ABET Student Outcomes. The critical/design thinking competencies are assessed through specially designed assignments in the Civil Engineering courses and with the help of a tailored critical thinking process based on an established problem-solving framework shown in Figure 2.4 and discussed later. These assignments incorporate the ASCE BOK3 foundations of critical thinking shown in Table 2.3. Emphasis is placed on the development of the appropriate level of critical thinking throughout the 4-year curriculum. The faculty accentuate that improved critical thinking skills result in increased ability to identify and understand problems and develop appropriate solutions. With higher levels of critical thinking, students will significantly improve in their analysis of complex engineering problems and produce higher quality solutions.

Civil Engineering is a practical profession that mostly involves the design and construction of various types of infrastructure. As such, critical thinking development within USCGA civil engineering program is predominately addressed through the design process. The design process has been successfully infused in the civil engineering curriculum through progressive and consistent integration of key design principles throughout the 4 years of education. This has been accomplished without adding new courses to the curriculum. Providing exposure to design is one of the requirements civil engineering programs must meet for ABET accreditation. ABET requires that students are "prepared for engineering practice through a curriculum culminating in a major design experience based on the knowledge and skills acquired in earlier course work and incorporating appropriate engineering standards and multiple realistic constraints" (ABET, 2018). Students are introduced to a problem-solving framework in the freshman year that serves as the foundation for further instruction in the design process in subsequent years. Several upper-level courses are structured to include project-based and cooperative aspects that promote learning at higher cognitive levels of Bloom's Taxonomy. A summary of how the various design or critical thinking aspects are progressively addressed in the civil engineering curriculum is presented in Table 2.4.

TABLE 2.4

Critical Thinking and Design Competency Addressed by Academic Year

Academic Year	Civil Engineering Course(s)	Design Activity/Critical Thinking Exercise	Bloom's Cognitive Level of Learning
Freshman	Statics and Engineering Design (1116) Engineering Mechanics (1118)	**Design a truss bridge/Failure case studies.** Define problem statement, identify and select appropriate analysis method, perform calculations, verify completeness of solution.	Knowledge and comprehension
Sophomore	Mechanics of Materials (1206)	**Design and perform an experiment. Compare test results with theory.** Define problem statement, identify and select appropriate analysis method, perform analytical and test calculations, verify completeness of solution.	Knowledge, comprehension, and application
Junior	Materials for Civil Engineers (1302)	**Determine Portland cement concrete mix and Asphalt concrete mix ingredients. Pavement thickness design.** Define problem statement, research problem, identify and select appropriate analysis/design method(s), investigate alternative solutions, perform design calculations, verify completeness of solution, prepare design documents.	Knowledge, comprehension, application, and analysis
	Soil Mechanics (1304) Steel Design (1313) Environmental Engineering II (1407)	**Investigate slope stability, complete technical paper.** **Design a steel truss pedestrian bridge.** **Design sanitary and sewer systems.** Define problem statement, research problem/technical, identify and select analysis method, use of software package, evaluation and verification of solution.	Knowledge, comprehension, application and analysis, synthesis and evaluation
Senior	Geotechnical Engineering Design (1404) Reinforced Concrete Design (1411) Civil Engineering Design (1402)	**Design of several geotechnical support structures** **Multi-story building design** **Capstone project-objectives dependent on nature of project** Define problem statement, site visit, research problem, identify and select appropriate analysis/design method(s), investigate alternative solutions, perform design calculations, select best alternative, verify completeness of solution, prepare design documents, assess impact of selected solution.	Knowledge, comprehension, application and analysis, synthesis and evaluation

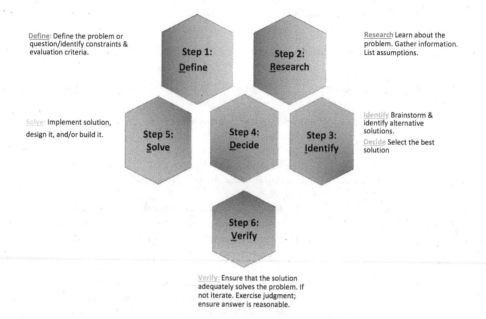

FIGURE 2.3   USCGA "DRIDS-V" problem-solving framework.

From freshman to junior year, students are exposed to various aspects of critical thinking and problem-solving in multiple courses. During the freshman year, engineering students are introduced to the problem-solving framework "DRIDS-V" (**D**efine, **R**esearch, **I**dentify, **D**ecide, **S**olve and **V**erify) developed by engineering faculty at USCGA. This problem-solving framework is used to improve the design thinking and critical thinking of students by integrating it into upper-level courses in the civil engineering curriculum. This approach is a step-by-step problem-solving technique that employs critical thinking in support of the general design process. Conceptually, this is an iterative process where steps 2–6 could loop back to each previous step. A detailed schematic of the "DRIDS-V" process is shown in Figure 2.3.

Prior to the fall of 2017, students in the freshman year were assigned a three-part truss bridge design project in the Static and Engineering Design course (1116). Each part of the project provided students with basic exposure to lower cognitive levels of Bloom's Taxonomy as indicated in Table 2.4. This bridge project was structured to mostly provide initial engineering problem-solving with exposure to engineering design concepts. Students' feedback was always positive, and they typically enjoyed working in groups to design, build, and test the 3D truss system to meet the specified requirements. This project motivated students to get hands-on design experience and better comprehend and apply the various principles covered in the course. This project also helped students improve their overall grade and performance in the course due to the connection made between theory and real-life projects.

The core curriculum at USCGA was revised in 2017 that resulted in some changes in the Static course. One of the changes made in the Statics course was to replace the bridge project with three failure case studies in a revised freshman Engineering

Mechanics – Statics (1118) course. Two of these case studies are selected and assigned by the instructors. Students are given the freedom to select the third failure case study based on their interests. Students are required to research these failure cases and give a presentation to the class that highlights the connections between the theory and practical aspects of the failure. Researching these case studies enables students to appreciate the practical applications of the principles and concepts covered in the course.

The concepts of the DRIDS-V problem-solving framework are expanded upon and reinforced during the sophomore, junior, and senior academic years as further preparation for engineering practice. Appropriate assessment tools are used to foster the progression from lower to higher levels of learning within Bloom's Taxonomy. Sophomores and juniors gain additional instruction and exposure to the design process in multiple courses as previously discussed in Table 2.4. Students are encouraged to develop and practice design and critical thinking skills by exposing them to practical applications of concepts concurrently with the appropriate theory. This has been successfully accomplished progressively with the use of project-based assignments or case studies in engineering courses. The framework used in upper-level courses is similar to the foundation of design thinking proposed in the ASCE-BOK3 outcome that is summarized in Table 2.5. Some of the ASCE BOK3 design thinking competencies shown in Table 2.5 are expected to be developed through mentorship and not necessarily during formal undergraduate education.

TABLE 2.5

ASCE-BOK3 Foundation for Design Thinking

Cognitive Domain Level of Achievement	Demonstrated Ability	Typical Pathway
Remember (remember previously learned material)	**Define** engineering design and the engineering design process.	Undergraduate education
Comprehend (grasp the meaning of learned material)	**Explain** engineering design and the engineering design process.	Undergraduate education
Apply (use learned material in new and concrete situations)	**Apply** the engineering design process to a given set of requirements and constraints to solve a complex civil engineering problem.	Undergraduate education
Analyze (break down learned material into its components parts so that its organizational structure may be understood)	**Analyze** a complex civil engineering project to determine design requirements and constraints.	Mentored experience
Synthesize (put learned material together to form a new whole)	**Develop** an appropriate design alternative for a complex civil engineering project that considers realistic requirements and constraints.	Mentored experience
Evaluate (judge the value of learned material for a given purpose)	**Evaluate** design alternatives for a complex civil engineering project for compliance with customary standards of practice, user and project needs, and relevant constraints.	Mentioned experience

Source: ASCE-BOK3- 2019, reproduced with permission from ASCE.

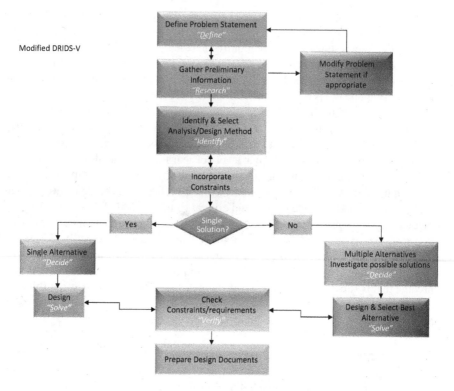

FIGURE 2.4   Advanced problem-solving/design process framework.

During the senior year, detailed instruction and substantial number of projects are implemented to further enhance critical thinking and the acquired knowledge of the design process. Design concepts are addressed in more detail in the three senior-level courses shown in Table 2.4. The "DRIDS-V" framework is expanded to include additional key design components. The expanded version of "DRIDS-V" (*advanced DRIDS-V*) is illustrated in Figure 2.4. The additional components introduced in the advanced DRIDS-V framework include:

- the option to modify the problem statement based on new information gathered
- multiple practical and more real-life constraints
- incorporation of additional constraints and limitations that may arise from the selected analysis method
- investigation of multiple solutions and identification of a "best" alternative
- a more detailed "verify" step that involves appropriately checking constraints, requirements, and specifications
- preparation of detailed design documents

The framework shown in Figure 2.4 mimics the engineering design process and is widely used in conjunction with Bloom's Taxonomy in three senior-level courses.

A summary of the learning and critical thinking outcomes for one of the courses (*Geotechnical Engineering Design*) is shown in Table 2.6.

In this project-based geotechnical engineering design course, students complete a series of open-ended design projects that are structured to balance the need for fundamental engineering instruction with an infusion of skills required for engineering practice. Students are required to incorporate engineering standards, develop alternative solutions, and consider economic, environmental, and other constraints as part of the design process using the advanced "DRIDS-V" framework. The pedagogical approach used in the course balances the need to cover basic concepts and design principles and spending time discussing and working on the projects. The projects are designed to progressively lead students through several cognitive levels of Bloom's Taxonomy. Details of the projects and the learning outcomes are also presented in Table 2.6. In project one, for example, students are required to remember material learned in the prerequisite *Soil Mechanics* course and apply them in the design process for a retaining wall. By the completion of the fourth project, cognitive learning is expected to have occurred at all six (knowledge, comprehension, application, analysis, synthesis, and evaluation) levels of Bloom's Taxonomy.

2.5.1 CASE STUDIES

Case studies were also progressively integrated starting in the freshman year through the upper-level courses. These cases are used as practical examples to reinforce technical concepts as well as foster professional development in ethics and life-long learning. Students are expected to apply knowledge they have acquired from their engineering and general education courses when reviewing these cases. The case studies were integrated into the courses with the following objectives:

- Provide background for specific engineering issues and associated principles
- Develop students' ability to identify and formulate problem statements
- Provide illustrations or practical examples of engineering concepts and principles
- Develop students' critical thinking and communications skills
- Promote life-long learning and professional development

In general, case histories are used in most civil engineering courses to enhance the learning experience beyond the traditional classroom activities and expose students to the analysis and decisions encountered by practicing engineers. The case histories provide background for specific topics, problem formulation and identification, and serve as illustrative examples of engineering concepts and principles covered in these courses. They tie together technical aspects, ethical issues, and procedural issues and help students engage in a higher level of critical thinking. They are used strategically throughout the semester to provide students with additional opportunities to make the connection between theory and real-life applications, as well as to generate interest in open discussion of engineering problems.

Examples of the implementation of case studies in three upper-level courses, *Geotechnical Engineering Design*, *Reinforced Concrete Design*, and *Water Resource Engineering*, are summarized in Table 2.7. As previously discussed, the

TABLE 2.6
Learning and Critical Thinking Outcomes in Geotechnical Engineering Design Projects

Project Title	Project Description	Learning and Critical Thinking Outcomes
Project 1: Retaining Wall Design	A retaining wall at a Coast Guard facility needs to be rehabilitated or replaced. The main objective is to design a new reinforced concrete retaining structure. Students are required to select suitable dimensions for the wall, evaluate the stability of wall by checking factors of safety against overturning, sliding, and bearing capacity failure. The ultimate bearing capacity of the soil must be determined to evaluate the capacity. Students must determine/calculate the amount of reinforcement for the stem, heel, and toe (as per ACI 318), sketch detailed construction drawing showing the location of the reinforcement. They must also make recommendation to mitigate the problems that caused the deterioration of the existing wall as well to minimize impact on the Coast Guard missions and operational readiness during construction.	Apply the theory of lateral stresses. Identify the effect of soil profile and surface on lateral stresses. Determine loading on wall due to active and passive lateral forces. Design a reinforced concrete retaining wall to all stability requirements. Determine/calculate the amount and size of reinforcement required. Estimate the construction cost of the project. Apply ACI 318 reinforced concrete design code. *Remembering previously learned materials in Soil Mechanics and making the connection to their relevance in the design process. Application, analysis, and evaluation through making judgment on causes of deterioration, choosing appropriate parameters to complete design, breaking down design problem into components.*
Project 2: Bulkhead Design	Students are required to apply the theory of lateral pressures to design a retaining bulkhead structure for a waterfront facility for mooring Coast Guard cutters and small boats. Due to tidal fluctuations, reasonable assumptions must be made on the height of the water level to be used in the design. Furthermore, the navigational draft of the largest cutter to be moored must be considered and a dredging depth recommended if required. Some constraints are placed on the soil and environmental conditions. Design must meet stability and other requirements.	Apply the theory of lateral stresses. Identify the effect of soil profile and surface on lateral stresses. Determine net loading on bulkhead due to active and passive lateral forces. Design a steel sheet-pile bulkhead including selection of suitable sheet-pile section to meet allowable stress requirements. Design an anchor or tie back to meet stability requirements. Familiarization with design codes. *Higher level of critical thinking and application of knowledge to complete design process. Evaluation through making judgment based on constraints.*

(Continued)

TABLE 2.6 (Continued)
Learning and Critical Thinking Outcomes in Geotechnical Engineering Design Projects

Project Title	Project Description	Learning and Critical Thinking Outcomes
Project 3: Pile Foundation Design	Students are provided information on the structural load, soil profile, and layout for a proposed bridge, and they are required to design a pile foundation system to support the superstructure of the bridge. The design includes the determination of the bearing capacity of pile group system, settlement calculations, and a comparison between single pile and pile-group actions. Students need to demonstrate their understanding of the design process, by selecting suitable analysis methods based on the soil type. Students are also required to evaluate and select the "best" option based on several constraints. Students must compare the results from their hand calculations with those from a pile design software.	Identify soil design parameters. Apply different static formulas to determine pile capacity. Identify the effect of soil profile and properties on capacity. Understand the importance of load tests. Design single pile and pile groups to meet vertical loading requirements. Determine settlement of single pile and pile group. Familiarization with methods of constructing drilled shafts. Compare results from hand and software calculations. *Application, analysis, and evaluation through choosing correct parameters to complete design, breaking down design problem into components. Synthesis by comparing design options and writing a comprehensive report.*
Project 4: Design of Shallow Foundations	The foundation for a new facility requires the use of a mat foundation system for the main building, and isolated footing for a walkway. The building layout is part of a combined project with the Reinforced Concrete Design Course. The structural analysis to determine the foundation loads shall be completed as part of the Reinforced Concrete Design Course. Students are required to design the mat foundation using the loading from the structural analysis using the rigid closed form solutions and one isolated footing ensuring that ACI standards are met. Information on the soil properties are provided; students are also required to determine whether settlement is within acceptable limits as well as check other relevant factors of safety. Students must compare the results from their hand calculations with those from a mat foundation design software.	Select design parameter based on soil properties. Understand the practical importance of flexural principles and apply them. Understand the interaction between soil and foundation. Analyze and determine soil-bearing capacity. Design a mat foundation. Design an isolated footing. Use and apply design codes. Compare results from hand and software calculations. *All six levels of Bloom's Taxonomy in cognitive domain. High level of critical thinking, making the connection with practical applications of knowledge gained in other civil engineering courses. Development of some professional judgment.*

TABLE 2.7
Examples of Case Histories

Case History	Problem Description/Statement	Learning Objective
	Geotechnical Engineering Design	
University building expansion, NJ	Dewatering and excavation support. Several challenges including underground utilities and limited access.	Addressing construction management issue and selection of suitable equipment.
Leaning tower of Pisa, Italy	Inadequate soil investigation, effects of differential settlement, selection of suitable soil improvement and stabilization techniques.	Understanding of soil-structure interaction, importance of thorough site investigation and selection of suitable foundation system.
Schoharie Creek Bridge failure, NY	Scour, inadequate bridge inspection and maintenance.	Importance of investigation and proper maintenance of bridges. Importance of understanding hydraulic forces on foundations.
	Reinforced Concrete Design	
Cocoa Beach five-story cast-in-place concrete building collapse, FL	Inadequate design, poor construction, and no quality control.	Addressing key construction issues including severe cracking and excessive deflection, proper rebar placements, punching shear strength, and lack of formwork design.
Skyline Center at Bailey's Crossroad high rise reinforced concrete building, VA	Catastrophic collapse of a high rise building due to no quality control and lack of formwork plan.	Importance of proper quality control related to punching shear, formwork (shoring and re-shoring slabs), and ensure proper concrete strength before formwork removal.

(Continued)

TABLE 2.7 (Continued)
Examples of Case Histories

Case History	Problem Description/Statement	Learning Objective
Failure during and after construction	Learning from the past failures during and after construction. Few cases reviewed in furthering the understanding and awareness needed to prevent failures. Consider meeting all the stages of checks and controls in the design, estimating, detailing, field supervision, and construction through which the project must go through in order to minimize the chance of failure.	Identify causes of failure. Appreciate the need for complete and updated construction drawings. Understand importance of good communications.
	Water Resources Engineering (Elective Course)	
Flood control reservoir operation during the Great Midwest Flood of 1993	Following the Great Midwest Flood of 1993, concern was voiced that the US Army Corps of Engineers did not operate flood control reservoirs in an optimal manner therefore contributed to the damage.	Understand reservoir operating plans particularly for flood control; evaluate (in hindsight) the effectiveness of reservoir operations during an extreme flood event.
Combined Sewer Outflow (CSO) Case study	In the Rouge River Watershed, MI, water quality was improved through CSO reduction.	Practical understanding of how CSOs are reduced and further explanation of the detrimental effects of CSOs.
Three Gorges Dam China	Case study provides an overview of challenges and decisions that were faced during construction.	Understand the complexity of the decisions that had to be made during the construction of this dam
Watershed Management on Groundwater and Irrigation Potential in India	An increase in population has led to increased water demand for irrigation and agricultural purposes in semi-arid and rural areas of India.	Understand the impact of watershed management on the groundwater and irrigation potential of a watershed facing drought-like conditions.

Geotechnical Engineering course is project-based with several open-ended problems that require students to make decisions and develop alternatives. By balancing the need for fundamental engineering instruction with that of cooperative learning, the development of problem-solving skills required for engineering practice is promoted in this course. The geotechnical engineering case histories infused into the course include real-life design, construction, and legislative disputes.

The project-based Reinforced Concrete Design course has been a required course where students learn the basic analysis and design of several structural components and apply these concepts in a semester project. The use of case histories and a multi-story building project advances students' comprehension of reinforced concrete design principles. Several case histories are selected from published literature and strategically introduced throughout the semester. These cases are used to either introduce a new topic or to highlight critical design and construction considerations during the design process. This approach enhances the learning environment by engaging students in discussions about what went wrong and what should be done to avoid similar mistakes in the design and construction of structures.

The Water Resources Engineering course is offered as an elective in the civil engineering curriculum. This course offers a basic introduction to water resources engineering and exposes civil engineers to a broad range of topics relevant to the field of water resources. Course topics include surface and groundwater hydrology, rainfall-runoff analysis, reservoir and river routing, probability and frequency analysis, computer modeling, water excess management/control, and watershed management. Case histories are used in this course to introduce students to issues surrounding water resources and to challenge them to recognize that engineering decisions are not solely based on hand calculations or computer simulation.

Suitable case histories are extracted from a wide variety of published resources (Delatte et al., 2012; Delatte, 2009; Kaminetzky, 1991; Watkins, 2013; Johnson et al., 2013; Gleick, 2008–2009, and Rouge River project, 2014). Examples of some of the case studies students received ahead of time in the three courses are included in Table 2.7.

Students are required to review each case and address key aspects in their write-ups and class discussions. Students are provided with the following guidelines that address some of the key aspects that must be considered while reviewing each case:

a. *Review the Case Content* – Provide a summary of the case.
b. *Identification of the Problem* – Identify key issues of the case and develop a problem statement.
c. *Collection of Relevant Information* – In general, no additional literature review or search for new information is formally required. However, students are expected to make the connection with already acquired knowledge and evaluate if any additional information is required to fully assess the case. In a few of these cases, instructors purposefully withheld key information to get students more engaged in this step.
d. *Development of Alternatives* – Students were encouraged to come up with alternative ways of solving the problem. Open classroom discussions facilitate the exchange of ideas, and students are encouraged to question the validity of each proposed solution.

e. *Selection of a Course of Action* – Having completed the first four steps, students are then encouraged to identify suitable approaches or solutions they think is most appropriate to address the issues based on the available information. Students are asked to select or identify the most appropriate course of action to resolve the problem. The cases presented may not have a straightforward solution.

f. *Recommendation of an Implementation Plan* – If a solution can be identified, students are required to propose an implementation plan that addresses any constraint defined in the problem statement.

2.5.2 CROSS-COURSE-COLLABORATION

Another strategy is the use of a comprehensive multi-story building project and common grading rubric in multiple courses. This "cross-course" collaboration involves the analysis and design of a multi-story building project. This project was developed in collaboration with several instructors and include components from structural design, geotechnical engineering, and construction management. The project was integrated into three senior-level design courses (Reinforced Concrete Design, Geotechnical Engineering Design, and Construction Project Management) in 2012 to mimic real-life projects and prepare students for the senior design capstone project. Each course requires students to perform the analysis and design of components of the project related to that course (Appendix 2). For example, the design of the mat foundation and retaining wall including slope stability analysis is completed in Geotechnical Engineering Design; the analysis and design of the floor slabs, beams, and columns including design drawings are completed in Reinforced Concrete Design; cost analysis and construction schedule are completed in Construction Project Management. The course instructors make extensive efforts to coordinate the activities and to guide students through the design process. This project lends itself well to continued development and integration with other courses where students can draw upon their knowledge of foundation design (gained through the Geotechnical Engineering Design course), designing various reinforced concrete elements (gained through Reinforced Concrete Design course), as well as their knowledge of constructability issues, construction cost estimating, and life-cycle cost analysis (gained through the Construction Project Management course).

A shared grading rubric is used concurrently in the *Geotechnical Engineering Design* and *Reinforced Concrete Design* courses to assess students' design competencies and communication skills. Students receive a copy of the grading rubric (details in Chapter 6) on the first day of class and are encouraged to consult it when completing each phase of the project (Appendix 5). The main categories assessed were grouped as follows:

- Definition of the problem statement, identification of constraints, and scope of the project.
- Research or gathering of information from relevant sources such as standard codes and specifications; identification of objectives.
- Identification and selection of appropriate analysis, design method(s), and design parameters.
- Consideration of practical constraints and constructability issues.

- Application of analysis and design methods.
- Preparation of engineering sketches and professional design or construction drawings.
- Submission of a professional engineering report detailing all the analysis and design calculations.

2.5.3 DESIGN-BUILD-TEST PROJECT

The Reinforced Concrete Design course also incorporates a "design-build-test" of full-scale reinforced concrete beams. The objective of this exercise is to investigate different modes of failure (shear, compression, transition between compression and tension, and ductile). Students (working in teams of 4–5) are required to select a mode failure, design for that failure mode, build, and test the beam. Each concrete beam is assumed to have constant material properties, constant span length of 10.5 ft, cross-sectional area of 100 in², and similar two-point loading conditions. Each team designs the necessary reinforcement to demonstrate the assigned failure mechanism and submit a technical report. Students build the reinforcement cages (Figure 2.5) based on their design and use ready-mix concrete to cast their beams. The ready-mixed concrete of a specified strength is delivered by truck. Students work together to properly place the concrete in the formwork, consolidate it, finish the surface, and prepare standard concrete cylinders for compression tests. Each team must ensure proper curing of the concrete and conduct compression tests at 7, 14, and 28 days. The 14-day compressive strength is used to calculate the expected failure load prior to actual testing of the beams using the full-scale testing apparatus shown in Figure 2.6.

The testing apparatus is used to apply two concentrated point loads at the 1/3 points along the length of the beams. Typical total failure loads range from 50,000 to 120,000 pounds. Concrete strain gages are embedded in the compression zone about

FIGURE 2.5   Typical rebar cages being placed in formwork.

FIGURE 2.6 Reinforced concrete beam testing setup.

½ inches from the top surface in each beam to collect strain data during loading. The recorded strain and deflection data are compared with the theoretical data to evaluate the accuracy of the design. During testing, students can see first-hand how constructability issues impact the performance and failure modes of the beams. They also gain a better understanding and appreciation of the various failure modes and factors of safety incorporated in the American Concrete Institute (ACI-318) building code. Each team submits a final technical report that includes iterations of flexural and shear design, analysis of the constructed beam with expected failure load, test results, conclusions, and recommendations. This project enables students to visualize the concepts and provide a hands-on opportunity. Students have fun completing all aspects of the project especially testing the various failure mechanisms. Cognitive learning typically occurs at all six levels of Bloom's Taxonomy by the completion of this project.

2.5.4 CAPSTONE PROJECT DESIGN

The completion of a capstone project during the senior year provides an excellent avenue for students to demonstrate and improve their critical thinking skills. The capstone experience is designed to provide a forum to practice the art of engineering under conditions encountered in Coast Guard and civilian engineering practices. Students work in teams of three to five and they oversee all aspects of the project. Course coordinator, faculty advisor, and sponsors serve as consultants to the team(s). Students are provided specific guidelines and project management techniques to help them produce professional results in a format suitable for engineering practice. Each team is required to break down their project into tasks, including a site visit to investigate the project, mandatory discussions with stakeholders, development of an acceptable solution, cost analysis, and the applicable consideration of environmental and social factors related to the design alternatives. Emphasis is placed on following the Coast Guard guidelines (or other military guidelines), relevant codes, and specifications. Guidelines and minimum deliverables for each project team include:

- Site visit report to verify the problem and/or refine project scope,
- weekly progress meetings with advisors followed by documented meeting minutes that are disseminated to the project team, the advisor, course coordinator, and the stakeholders,
- a final comprehensive project report,
- a project notebook that demonstrates their progress throughout the semester (provided as a reference to the sponsors),
- two poster presentations of project status throughout the semester,
- a final public presentation to all stakeholders of the capstone project.

Projects vary in complexity, but they all provide students with real exposure to the design, planning, and management of actual civil engineering projects. The successful completion of the projects involves extensive critical and design thinking at the highest level of Bloom's cognitive learning.

2.6   CLOSING THOUGHTS

The focus of this chapter was on the importance of critical and design thinking in undergraduate engineering education. Critical thinking is essential in problem-solving and in the development of appropriate solutions. Teaching, learning, and assessment of critical thinking is challenging, but this must be continuously and consistently done at the undergraduate level. To facilitate the development of critical thinking skills, instructors must develop specially designed assignments or projects that gradually, over 4 years, promote the advancement of cognitive thinking. Teaching students to think critically requires more than simply providing them with facts, theories, and techniques. Creating frameworks or perspectives for critical thinking takes time, patience, and the intentional design of classroom exercises and assignments that guide students to practice critical thinking sequentially throughout their undergraduate studies. An approach to critical thinking instruction that is appropriate for undergraduate students can be based on the conceptualization of critical thinking that incorporates the 21st-Century Bloom's Taxonomy framework. These six domains of critical thinking and reasoning must be sequentially used to foster independent and critical thought.

The connection between critical thinking and design thinking was also discussed, and examples on how the development of these skills is addressed in the civil engineering curriculum at the United States Coast Guard Academy are presented. Examples on how the Bloom's Taxonomy model was adopted to provide a foundation in developing classroom materials and assessment instruments at the appropriate undergraduate level were briefly discussed with more details presented in Chapter 6. Design, case study, and project-based activities were progressively infused into the curriculum to help students perform at higher levels of cognitive learning. With higher levels of critical thinking, students significantly improve their ability to analyze complex engineering problems and produce higher quality solutions. Emphasis should be placed on student-focused teaching using a general problem-solving framework, project-based learning, and exposure to professional practice.

REFERENCES

ABET. (2018). *Criteria for Accrediting Engineering Programs.* ABET Engineering Accreditation Commission, Baltimore, MD.

Adair, D. and Jaeger, M. (2016). "Incorporating critical thinking into an engineering under-graduate learning environment." *International Journal of Higher Education*, Vol. 5, No. 2, pp. 23–39.

Ahern, A., Dominguez, C., McNally, C., O'Sullivan, J.J. and Pedrosa, D. (2019). "A literature review of critical thinking in engineering education." *Studies in Higher Education*, Vol. 44, No. 5, pp. 816–828.

Ahern, A., O'Connor, T., McRuairc, G., McNamara, M. and O'Donnell, D. (2012). "Critical thinking in the university curriculum – the impact on engineering education." *European Journal of Engineering Education*, Taylor & Francis, Vol. 37, No. 2, pp. 125–132.

American Society of Civil Engineers (ASCE – BOK3). (2019). *Civil Engineering Body of Knowledge, Preparing the Future Civil Engineer*, Third Edition, ASCE Press, Reston, VA.

Anderson, L.W. and D. Krathwohl (Eds.) (2001). *A Taxonomy for Learning, Teaching and Assessing: a Revision of Bloom's Taxonomy of Educational Objectives.* Longman, New York.

Bloom, B.S. (Ed.). Engelhart, M.D., Furst, E.J., Hill, W.H. and Krathwohl, D.R. (1956). *Taxonomy of Educational Objectives, Handbook I: The Cognitive Domain.* David McKay Co Inc, New York.

Cotter, E.M. and Tally, C.S. (2009). "Do critical thinking exercises improve critical thinking skills?" *Educational Research Quarterly*, Vol. 33, No. 2, pp. 3–14.

Delatte, N. (2009). *Beyond Failure: Forensic Case Studies for Civil Engineers*, ASCE Press, Reston, VA.

Delatte, N., Bosela, P. and Bagaka's, J. (2012). "Implementation and assessment of failure case studies in the engineering curriculum." *ASCE Forensic Engineering Conference*, ASCE Press, 2012.

Elder, L. and Paul, R. (2010). *Critical Thinking Development: A Stage Theory.* The Foundation for Critical Thinking Press, Tomales, CA. www.criticalthinking.org.

Facione, P. (1998). *Critical Thinking: A statement of Expert Consensus for Purpose of Educational Assessment and Instruction.* The Delphi Report, Insight Assessment and the California Academic Press.

Fitzgerald, M. (2000). "Critical thinking 101: the basics of evaluating information." *Knowledge Quest*, Vol. 29, No. 2, pp. 13–20, 22–24.

Gleick, P.H. (2009) "Three gorges dam project, Yangtze River, China." *The World's Water* 2008–2009, pp. 139–150.

Gude, V.G. and Truax, D.D. (2015). "Importance of critical thinking in environmental engi-neering." *ASEE Southeast Section Conference*, University of Florida, Gainesville, FL, 12–14 April.

Johnson, J.N., Govindaradjane, S. and Sundararajan, T. (2013). "Impact of watershed manage-ment on the groundwater and irrigation potential: a case study." *International Journal of Engineering and Innovative Technology(IJEIT)*. Vol 2, No. 8, pp. 42–4.

Johnson, R., Johnson, D. and Smith, K. (1998). *Active Learning: Cooperation in the College Classroom.* Interaction Book Company, Edina, MN.

Kaminetzky, D. (1991) *Design and Construction Failures – Lessons from Forensic Investigations.* New York, NY, McGraw-Hill, Inc.

Miller, R.K. and Linder, B. (2015). *Is Design Thinking the New Liberal Arts of Education?* Olin College, Needham, MA.

Paul, R. and Elder, L. (2009). *Critical Thinking: Concepts and Tools.* The Foundation for Critical Thinking Press, Tomales, CA.

Paul, R., Niewoehner, R. and Elder, L. (2013). *Engineering Reasoning based on Critical Thinking Concepts & Tools.* The Foundation for Critical Thinking Press, Tomales, CA.

Rouge River Project. Report to congress on implementation and enforcement of the CSO control policy. http://www.wcdoe.org/rougeriver/. Accessed 31 March 2014.

Scriven, M. and Paul, R. (1987). Critical thinking as defined by the national council for excellence in critical thinking. *8th Annual International Conference on Critical Thinking and Education Reform* online: http://www.criticalthinking.org/pages/defining-critical-thinking/766.

Watkins, D.W. (2013). *Water Resources Systems Analysis through Case Studies: Data and Models for Decision Making.* ASCE Press, Reston, VA.

3 Development of Leadership Skills

3.1 INTRODUCTION

There is an ongoing worldwide need for undergraduate engineering education to provide adequate balance between academics, professional skills, industrial relevance, and global perspectives. The influence of a global economy now requires engineers to not only be technically competent but to also have a diverse background, be sensitive to the needs of other cultures, be ethical, and demonstrate good leadership qualities. For new engineers to be successful, they must be able to function in this increasingly complex global market with adequate awareness of engineering needs and practices and acquire professional skills to lead across cultures. The approaches to developing a global perspective and critical thinking were discussed in Chapters 1 and 2.

There has been much concern in the civil engineering industry about the development of leadership skills and the abilities of graduates as this aspect is seldom adequately addressed in traditional engineering programs. Typically, engineers are not routinely prepared to lead groups of people and organizations through academic training or work assignments early in their careers. The civil engineering industry, professional and academic institutions understand the importance of developing leadership skills in the success of engineering graduates. The National Academy of Engineering in one of its publications, *The Engineer of 2020: Visions of Engineering in the New Century (2004)*, identified the importance of having engineers be knowledgeable in topics such as principles of leadership and management, ethical standards, cultural diversity, professional resiliency, global impacts, and cost–benefit constraints to ensure the future development of the professional engineering practice. The American Society of Civil Engineers (ASCE), in response to feedback from practicing civil engineering professionals, articulated new standards for civil engineering programs to meet in regard to these skills that are included in the professional component of the ASCE Body of Knowledge (BOK3, 2019). This professional component includes proficiency in communications, public policy, business and public administration, globalization, leadership, teamwork, attitudes, life-long learning, as well as professional and ethical responsibilities. The Engineering Accreditation Board (Accreditation Board for Engineering and Technology, ABET) also places emphasis on the successful preparation of undergraduates who are equipped with adequate technical and professional skills needed for the practice of engineering. Two decades ago, ABET introduced EC2000 "criterion 3 Student Outcomes a-k" which was a major departure from the methods used to accredit engineering programs in the United States of America (USA). Engineering programs adopted the assessment and accountability in preparing engineering students to practice in the engineering industry. These student outcomes listed under ABET's criterion 3 were

DOI: 10.1201/9781003280057-3

modified and adopted in 2018 as "Students Outcomes 1–7" that are now used to assess attributes for graduates entering the engineering profession. Therefore, it is the responsibility of academic institutions to ensure that students have ample opportunities to develop and improve both the technical and professional skills required for successful professional practice of engineering.

It has been envisioned that an effective engineering leader develops positive professional relationships, respects others, communicates effectively, shows passion, embraces diversity, mentors and encourages others, builds multi-cultural teams to shape outcomes that meet established objectives, and is compassionate and sympathetic to the needs of others. Helping students develop such skills is not easy to accomplish in 4 years of undergraduate education. A good start is for academic institutions to make the development of professional attributes an important part of their missions. Opportunities for students to acquire leadership skills and experience can then be provided through coursework and extracurricular activities. Through coursework, students can be exposed to established principles of leadership and case studies, interact with successful leaders, and apply such concepts into practice by participating in extracurricular activities. The leadership competencies in these categories may include conflict management, equity and diversity, teamwork, cultural understanding, decision-making, problem-solving, vision development and implementation, creativity and innovation, technology management, financial management, strategic thinking, and entrepreneurship. This chapter discusses several models/approaches used in academia to develop leadership skills and provides case-study examples used at the U.S. Coast Guard Academy.

3.2   FRAMEWORK FOR LEADERSHIP

There are many ideas in the published literature on leadership development and successful implementations in engineering. For example, Skipper and Bell (2006) recognized that leadership is a complex subject that is impacted by variables such as skill requirements at different stages in someone's career, the varying role assumed by leaders within different organizations, and the impact of technology. They highlighted the leadership skills needed to be successful in the way they defined "leaders" as persons who recognize the need for and implement change, establish direction, align people, motivate and inspire, delegate as opposed to hoard power, communicate a vision of where the organization is headed, build diverse teams, share decision-making, mentor and coach subordinates, and demonstrate a high degree of integrity in their professional interactions.

Stogdill and Bass (1981) defined Leadership as the "...process of influencing the activities of an organizational group in the efforts toward goal setting and achievement." While Houghton and Neck (2002) defined the term self-leadership as "...a process of influencing oneself to achieve the self-direction and self-motivation necessary to behave and perform in desirable ways." Manz and Sims (2001) contend that individuals should first be trained in self-leadership so that they lead themselves before they start leading their organizations. Manz and Sims also argued that leadership skills must build upon organizational values and principles. Hawken (2007) suggested that leaders' action should be coined from a deep sense of commitment to their ideologies, codes, and values that include justice, equity, respect, honesty, integrity, stability, humility, creativity, flexibility, openness, diversity, and sense of community. Furthermore, Neck (1996) indicates that self-leadership is critical to organizational effectiveness as efficient self-leaders are more dynamic and less resistant to organizational change and hence more impactful on organizational productivity and technological creativity. As a process, Riggio et al. (2003) observed that leadership works to create productive human relationships through effective communication. Leadership is made up of strong oral and written communications, interpersonal skills, and the ability to manage problems and the task environment. Communication is considered a valued leadership skill where interaction with others allows leaders to engage in conversation and enables them to be attentive, perceptive, and responsive to others.

DeLisle (1999) defined leadership as a direct function of three elements of interpersonal effectiveness: (1) people's awareness of themselves and other people; (2) their ability to make decisions, solve problems, motivate others, and balance the tasks and relationships in an organization; and (3) the commitment to make hard decisions and face the risk of "doing the right thing." DeLisle suggested several traits engineers must develop to become effective leaders:

a. seek feedback and information about their interpersonal effectiveness,
b. develop understanding of how people are motivated and how they grow and develop in the organization,
c. develop and sustain conceptual flexibility and be comfortable with change and ambiguity,
d. hone technical problem-solving skills but develop an equally competent set of interpersonal skills related to communications.

Todd (1996) recommended that leaders need to acquire broad knowledge in engineering, public affairs, environmental law, finance, communications, and interpersonal skills. Todd also suggested that leaders must practice mentoring, must be willing to work with others on a multi-cultural team, and assist the nation in finding new approaches for rebuilding the infrastructure for enhanced quality of life and global competitiveness. Todd also recommended redesigning the organizational culture, developing organizational core values, and emphasizing how to lead, manage, and how to best create multidisciplinary teams.

Leadership has been considered as the most important source of advancing competitive advantage. Tarabishy et al. (2005) identified that effective organizations require leaders who can determine and deliver desired changes in structure, cultural climate, process, and governance within their organizations. Organizations need leaders who are equipped to embrace change and uncertainty as well as train their followers to face the constantly increasing changes and challenges in business and government environment (Northouse, 2007; Dodgson, 2011). Kouzes et al. (2010) described effective leaders are those who can transform their organizations into creative, effective, and productive enterprises by providing their organizations with the direction, vision, drive, motivation, and push to bring success. Kouzes and Posner (2011) also argued that leadership works when values are clarified, understood, and aligned with action. According to Wheatley and Frieze (2011), true leaders take action to create the world they want to see and encourage others to act precisely in the same manner.

Athreya et al. (2008) reported the establishment of the Engineering Leadership Program (ELP) at Iowa State University to develop engineers with the characteristics outlined by the National Academy of Engineering in the *Engineer of 2020*. The ELP Leadership Model identified four broad categories into which 19 leadership competencies are grouped:

1. leadership characteristics such as initiatives, integrity, analysis and judgment, communication, energy, and drive.
2. engaging others by building successful teams, developing others, coaching, teamwork, and leading through vision and values.
3. awareness and growth by developing engineering knowledge, general knowledge, cultural adaptability, and continuous learning.
4. demonstrating excellence through quality work, customer focus, innovation, professional impact, and planning.

Thayer (1988) presented an approach to the study of leadership consisting of the examination of leader characteristics, follower requirements and characteristics, and analysis of strategies on how to develop various skills so that leaders will effectively communicate, engage, build, trust, influence, and understand people and their motivations while leading their organizations. Kaiser et al. (2008) argued that leaders who can influence, plan, coordinate, guide, and decide, as these skills form a core competency of the military, are needed. Military organizations have been pioneers in the leadership field (Taylor and Rosenbach, 2005) delivering instruction on leadership that ranges from the entry level, into the officer corps, and even the flag officer level.

Academic institutions' integration of leadership theory with real-world practices has gained a fundamental role in demonstrating the importance of self-leadership (Prussia et al., 1998) and leadership education (Maellaro, 2013). Brown and Posner (2001) believe that leadership development is a learning process where students learn to lead both through appropriate classroom education with specifically designed learning objectives and real-life experiences (Arendt and Gregoire, 2005). Similarly, Day (2001) argues that leadership education must consist of the development of professional skills such as communication, critical thinking, and business competencies together with leadership growth where the personal integration of theory is mixed with practice and training.

3.3   LEADERSHIP DEVELOPMENT AT USCGA

The United States Coast Guard Academy (USCGA) has been educating future leaders for the U.S. Coast Guard (USCG) since 1876. The USCGA, located in New London, Connecticut, is the smallest of the United States federal military academies with a mission to "educate, train and develop leaders of character who are ethically, intellectually, and professionally ready to serve their country and humanity." An academy-wide culture of leadership is required to facilitate learning across academic disciplines, military training, and athletic training. This section presents an example of leadership development where students (cadets) are trained over 4 years (200-week journey) to become leaders through a formal leadership education that is incorporated into nine academic disciplines: Civil Engineering, Cyber Systems, Electrical Engineering, Mechanical Engineering, Naval Architecture and Marine Engineering, Government, Management, Operations Research and Computer Analysis, and Marine and Environmental Science. The unique trait of the leadership development at the USCGA is based on the development in four global stages: *leading self, leading others, leading performance and change*, and *leading the Coast Guard*. Leadership development is accomplished by exposing cadets to models of leadership via curriculum and integrating theory with practice through extracurricular and service-learning experiences to fuse academics, military training, and the athletic programs. During the 4 years, academic, military and athletic programs work together to prepare, guide, mentor, and monitor students' development and implementation of moral, ethical, leadership, and other professional skills expected of new graduates.

USCGA uses *the L.E.A.D. Program* for students' leadership education and training. The L.E.A.D. program was developed at USCGA, and this dynamic strategy is continually assessed and revised to improve its effectiveness to graduate competent USCG leaders (COMDTINST 5351.1). The various components of this model are as follows:

- *L (Learn from Theory)* – The "L" indicates that students *Learn from Theory*. Students are taught the foundational and advanced leadership models, as well as theories related to specific domains and their changing roles from followers as freshmen (4/c cadets) to leaders as seniors (1/c cadets).
- *E (Experience through Practice)* – The next stage in this model, "E," is that students also learn from *Experience through Practice*. From being a watch stander or member of the Regimental Staff, an athlete or captain of a sports team, and a leader in a group project or any academic coursework, students are given ample opportunities every year to practice leadership skills and learn from their experiences.
- *A (Analyze Using Reflection)* – Students are also expected to learn how to *Analyze using reflection*. With the assistance of military and academic faculty or on their own and driven by their own initiatives, students develop various intellectual pursuits that allow them to analyze and reflect. They often write professional leadership papers in Student Leadership Journal entries and write leadership term papers in several academic courses. For example, in some of the assignments such as *Identity Papers, 360 Feedback*, and *Leadership Defense* papers, students are required to reflect on opportunities that are designed to encourage self-development as leaders throughout all 4 years of their college education.
- *D (Deepen through Mentoring)* – It is not until senior year (1/c year) that students can *deepen their experiences through mentoring* during which they are exposed for the first time to mentoring through the L.E.A.D. Mentoring program. Mentoring from military staff and academic and athletic faculty provides students with leadership models which students take and apply

FIGURE 3.1   Schematic of USCGA's leadership development model.

through their experiences in the residential halls, classroom, sports field, or during any club activity. These frequent quality interactions are designed to provide cadets with guidance to be future officers and *leaders of character* serving the U.S. Coast Guard and the nation.

The leadership development that promotes the components of the L.E.A.D. program is applied in the four categories: *leading self, leading others, leading performance and change,* and *leading the Coast Guard* as illustrated in Figure 3.1. The entire model is centered around first learning how to progressively "leading self" before focusing on the other components. *Leading self* is essential to the successful development of a leader. Leaders must understand "self and one's own abilities" such as personality, values, and preferences while simultaneously recognizing their own potential. *Leading others* involves working with and influencing others to achieve shared goals. USCG members are required to interact with others to build positive professional relationships that serve as a foundation for the success of completing missions. *Leading performance and change* require USCG members to continuously face challenges in mission operations as challenges change overtime. These four categories consist of 28 competencies or skills shown in Table 3.1 that USCGA considers critical in developing effective leader. Students have ample opportunities to develop and practice these skills over their 4 years at USCGA.

TABLE 3.1

USCGA Leadership Development Framework/Competencies

Leadership Categories	Leadership Competencies
Leading self (Freshman year)	Accountability and responsibility, aligning values, followership, health and well-being, self-awareness, personal conduct, technical proficiency.
Leading others (Sophomore year to Senior year)	Effective communication, team building, influencing others, taking care of people, mentoring, respect for others, and diversity management.
Leading performance and change (Sophomore year to Senior year)	Conflict management, customer focus, decision-making and problem-solving, management and process improvement, vision development and implementation, creativity, and innovation.
Leading the Coast Guard (Senior year and beyond)	Financial management, technology management, human resource management, external awareness, political savvy, partnering entrepreneurship, stewardship, strategic thinking.

As previously discussed in Chapter 1, during their freshman year, the focus is predominately on the competencies detailed in the leading self-category. Freshman year is one of the most important stages of a student's academic life. This is when much of self-discovery, understanding of the academic system, as well as an early development of technical proficiency in the areas of study and learning take place. With each progressive year, the students shift their focus to the next set of competencies while still working on honing their skills in the previous categories. Details of these competencies will be discussed in the following section.

The initial and most vital component for the successful development as a leader is an understanding of self, including one's strengths and weaknesses. Within the "leading self" category are competencies such as personality, self-awareness, health and well-being, honor, devotion to duty, and technical proficiency as summarized in Table 3.1. Academic institutions can help students develop these or similar competencies by creating an environment that encourages self-discovery through diverse academic and non-academic activities. For example, freshman at USCGA starts their military training and leadership development during the summer prior to beginning of the academic year. The freshman begins as follower, assimilating into the rigors of military life while developing teamwork skills essential for success in the Coast Guard. The primary focus of freshman year is learning the Coast Guard and Academy cultures, while rapidly mastering an extensive amount of entry-level knowledge and balancing their time between academics, military, and athletics. During the sophomore year, students' transition into entry-level leadership by becoming role models for the new freshman class and internalizing key leadership principles highlighted in Table 3.1. During the junior year, students are responsible for mentoring and inspiring the underclass. They serve as trainers to the incoming freshman class during the summer and mentors during the academic year. Juniors take on more leadership roles within the Academy and are expected to handle themselves appropriately in more challenging situations and are accountable for themselves and their peers. Finally, during the senior year, students take on the leadership role for the Corps by filling roles as Regimental Staff Officers, Company Commanders, Department heads, and Division officers within the Corps of Cadets. Seniors lead the Corps (entire student body) by direction and example, serving as positive role models, while in military uniform and civilian attire. Since leadership development is a continuous process, there are overlaps between the categories listed in Table 3.1, and development opportunities are spread throughout the 4-year curriculum.

Other academic institutions can help students develop similar competencies by creating an environment that encourages self-discovery through diverse academic and non-academic activities. Effective time management is a critical skill every student must develop in order to adjust to college life, improve study habits, explore new opportunities, serve the community, mentor underclass, tutor, and start building a professional career. One suggestion is to consider expanding the typical orientation program into an "Introduction to Leadership" camp that would be structured to include some leadership training for students. This camp could focus on exposing incoming freshmen to some of the fundaments of leadership as well as the typical introduction to college life, time management, ethical behavior, and cultural expectations and responsibilities. The concept of orientation is not new to academic institutions in the USA. This model can be built upon and expanded to provide more community building and leadership development activities. Such summer orientation camp could be structured to build relationships between students, establish communities, experience fun activities, discuss core values and expectations, and highlight opportunities on campus or in the local communities.

To facilitate the development of the four leadership categories at USCGA, a triad approach that consists of three basic set of activities – academics, non-academics (that include military and athletics), and professional practice (include professional membership and professional development activities) as shown in Figure 3.2 and described as follows:

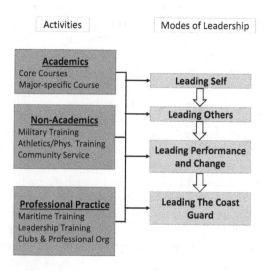

FIGURE 3.2   Leadership development category of activities.

- *Academic Activities* – this includes a core curriculum with a variety of courses that enable students to expand their horizon in addition to major specific technical or non-technical courses that foster the development of leadership skills. Opportunities for growth are progressively infused into the curricula.
- *Non-academic Activities* – these set of activities include military training, physical health and fitness, and community service.
  - "*Honor, Respect, and Devotion to Duty*" are very important core values mandated in the USCGA mission statement. Teaching of leadership principles within the cadet/officer training programs is reinforced through a combination of an interactive dialogue with their civilian and military instructors and by the co-curricular teaching activities and personal examples set by dedicated officers and faculty members. The Honor Code is enforced and administered through the Cadet Honor System. Upperclass cadets are responsible for the ongoing instruction of underclass in the principles and obligations of living under the Honor Code. Senior cadets who serve as Regimental Officers receive, investigate, and conduct Honor Council hearings of alleged violations of the Honor Code by members of the Corps of Cadets. During the summer, cadets have additional opportunities to apply their educational experience in USCG operations during internships, summer assignments on ships and at USCG units, or at USCGA while training incoming cadets. The competencies already developed are reinforced and applied including team building, communications, accountability, conflict management, and technical proficiency. It is important to understand the holistic approach to leadership development that each cadet is required to experience in order to grow into a competent Coast Guard leader. Over the 200-week (4 residential years) educational experience, students (cadets) go through numerous undertakings in academics, athletics, and military to *learn* and *reflect* on theories, *experience* leadership, and *reflect* upon their own leadership development and growth. As leader development requires personal interactions and practice, cadets experience the 200 weeks of leadership opportunities where they learn from failure, are expected to struggle to meet some competencies, and grow as a professional. Each academic year, cadets get the opportunities to take on new and distinctive roles with in the Corp of Cadet that further fosters their growth as leaders.
  - New USCG officers are operationally ready when physical preparedness and strengthening standards are met and accompanied by leading a healthy and fit lifestyle. The leadership program assumes that all USCGA cadets can improve their health status because fitness levels are dependent on a sound exercise program. Health fitness enhances the individual's capacity to handle the physical demands of the job, reduces the risk of injury and illness, assists in weight management, and helps individuals pursue physically active recreational activities. The responsibility for the physical-athletic development of cadets' rests primarily

with the Director for Athletics and the Commandant of Cadets. Together their staff provide a range of activities that shape both the bodies and physical skills of all cadets.

- During 4 years at the Academy, all cadets go through a set of training exercises, drills, physical fitness training, and sport activities. To achieve high levels of operational effectiveness and readiness, cadets are required to be physically fit and meet the standards of various physical fitness tests so that they are ready and able to perform their military duties. The USCGA Leadership Development program considers physical fitness as a critical element of the mental discipline and moral strength that can be acquired through participation in competitive sports when teamwork and leadership are practiced and developed.

- Cadets are required to develop expertise in a variety of sports and demonstrate tactical and technical competence by being able to explain and demonstrate athletic activities and know the best methods of preserving and performing them. Concepts such as fair play, teamwork, and sportsmanship are moral elements that result from understanding core values of honor and respect. Physical fitness is a vital component of cadet life and benefits cadets by instilling an appreciation for maintaining an active personal life-long fitness regime.

- Participation in non-academic activities is part of the USCGA culture where cadets are required to fit these extracurricular activities into their busy schedule. To learn how to lead others, cadets are also encouraged to participate in approved social clubs and community service. Throughout these activities, cadets practice teamwork, goal setting, diversity appreciation, tolerance, conflict management, and communication. The Academy recognizes the importance of community and service learning as they are vital for development and enhancement of cadets' leadership skills. Through community work and service learning, cadets apply their academic skills and knowledge to address real-life needs in their communities and become enriched in ethnically and culturally diverse environments. Cadets are required to complete at least 8 hours of community service each semester. This provides tremendous opportunities for cadets to engage with local communities and to explore practical ways of serving others. By working with local community leaders to solve community-based problems, cadets go through actual experience and active experimentation followed by reflection as they seek to achieve real objectives for the community. These hands-on opportunities followed by reflection provide cadets with deeper understanding of their self-leadership and other leadership skills. Another important area of measurable service learning is achieved through cadets' involvement in the social and residential life at USCGA. Cadets are exposed to self-leadership development and learning opportunities by living together, designing their own social interactions, and collaborating socially and professionally regularly with faculty, staff, coaches, and administrators outside the classroom.

They are also governed by a standard code for cadet conduct and academic integrity. The military, residential, and social environments are critical platforms of practical experience and active experimentation that are designed to bring together talented, engaged, and energetic cadets with various leadership abilities, interests, racial, socioeconomic, and ethnic backgrounds to promote self-leadership and the Coast Guard's code of conduct. Furthermore, during the summer, cadets have additional opportunities to apply their learning in the operational USCG, internships, and coastal sail or are at the Academy training incoming cadets.

- *Professional Practice* – maritime training, leadership training, and running clubs and professional organizations. Cadets, as members of the Corps of Cadets, have opportunities to take active leadership roles in the running of daily life on campus and managing other cadets. Throughout the process, cadets rise to the challenge of experiencing, developing, and practicing their leadership skills in the Coast Guard by training and mentoring other students, lead a student organization, active participation in sports, active participation in the community, and organize and run activities for the students at the Academy. Sophomores serve as mentors and tutors to the freshman, overseeing the new transition to military, college, and academy life. Juniors and seniors, in addition to leading the freshman and sophomores, play an active role in running the day-to-day operations of the Academy and the Corps of Cadets. By graduation, each student would have had several opportunities to serve in positions that involve leading other students thus developing and honing their competencies in this category. Students support self-governing organizations and clubs (professional or social) to promote and develop self-leadership training outside the academic classroom or military training. Faculty and staff serve as advisors, but these organizations are governed and managed by students. Faculty advisors ensure that students are motivated to promote the missions and activities of the organizations. Participation provides additional opportunities for cadets to practice "self-leadership" and "leading others" skills taking responsibility of managing the organizations. With each progressive year, students shift their focus to the next set of leadership competencies, while still sharpening their skills in the previous categories.

3.3.1 CASE STUDY: LEADERSHIP DEVELOPMENT IN THE CIVIL ENGINEERING CURRICULUM

Approximately 15% of the cadet corps graduates with a Civil Engineering degree. The USCGA's mission is: "to educate, train and develop leaders of character who are ethically, intellectually, and professionally prepared to serve their country and humanity." The civil engineering curriculum is broad and provides a solid background in the environmental, geotechnical, structural, and construction sub-fields of civil engineering. Graduates pursue several different career paths, and many of them serve in the Coast Guard as practicing civil engineers, pursue professional licensure, and attend graduate programs in Civil Engineering. Emphasis is placed on balancing theory and practice of engineering, so graduates are intellectually and professionally prepared to provide engineering services to the Coast Guard. Professional skills are particularly reinforced in the engineering courses through laboratory reports, technical papers, presentations, design projects, field trips, interaction with practitioners and Coast Guard officers, community outreach activities, and professional membership.

The academics component of leadership development shown in Figure 3.2 is supported by a strong "Core Curriculum" of science, math, engineering, professional studies, and humanities courses. This "Core" of 102 semester credits must be successfully completed by every student prior to graduation in addition to the major specific requirements for civil engineering of at least 34 credits. The wide variety of core coursework outside of engineering help to widen students' horizon, encourage self-criticism and discovery, and develop professional skills. Civil engineering students take at least 34 credit hours of non-technical core courses plus six additional credits of Health and Physical Education. Since leadership development is a continuous process, there are overlaps between the categories listed in Table 3.1, development and practice opportunities are spread throughout the curriculum.

In the present core curriculum, during the freshman year, students across all majors are required to take two courses: *Leaders in U.S. History* and *Macroeconomic Principles*. In the *Leaders in U.S. History* course, students survey major developments in U.S. History through the lens of key leaders where they evaluate different models of effective and ineffective leadership and management. They are also required to take *Macroeconomic Principles* to understand basic economic concepts as well as develop competency in financial management.

During the sophomore year, civil engineering students take two courses: "Leadership and Organizational Behavior" and "American Government." In the "Leadership and Organizational Behavior" *course*, students are exposed to fundamental leadership and management concepts such as values, personality, self-awareness, goal setting, working in teams, motivation, and setting a vision with particular emphasis on the practical leadership implications. The focus of the course is on leading self and leading others and how these components tie into the competencies listed in Table 3.1. In the *American Government* course students study the political process and the making of public policy, examine the framework of the U.S. democratic system, and explore topics on political parties, election processes, interest groups, and civil liberties as well as domestic and foreign policy including the policy making process and its consequences. Furthermore, students learn about *leading others* and understanding the working relationship with local, state, and federal government agencies. This course helps prepare civil engineers to understand the working relationship with local, state, and federal government agencies.

During the junior year, students take *Morals and Ethics* and *Criminal Justice* courses. As seniors, they continue their study of law in *Maritime Law Enforcement* course. In the *Morals and Ethics* course, students examine a variety of philosophical models regarding actions that can be considered right or wrong. By analyzing ethical and moral standards students reinforce their self-decision-making abilities and develop their own moral voice. In *Criminal Justice* and *Maritime Law Enforcement*, they are required to study the U.S. civilian and military criminal justice system and the legal issues associated with the Coast Guard's law enforcement mission in the maritime environment. In their senior year, while completing a capstone project, students are exposed to teamwork in research or project work where they practice leadership by managing group members, rotate chairmanship of weekly project meetings, complete deliverables, etc. that contribute to learning and understanding their best attributes of personal leadership style. During 4 years of undergraduate studies, students are required to develop and master *leadership abilities, personal and professional qualities, communication effectiveness, ability to acquire, integrate and expand knowledge*, and *critical thinking skills*. As previously mentioned, developing leaders is a continuous process, and there are overlaps between the four categories highlighted in Table 3.1. Throughout the process, students rise to the challenge of developing and practicing their leadership skills in the Coast Guard by training and mentoring other students, leading a student organization, actively participating in sports, being active in the community, and organizing and running activities for the Corps of Cadets and the Academy.

In addition to academics, students are challenged non-academically (militarily, athletically, and socially) through daily interactions with each other, USCGA faculty and staff, and through structured military and athletic training opportunities. With over 30 National Collegiate Athletic Association sanctioned teams and multiple nationally ranked sports club, students have numerous opportunities to develop and practice their leadership skills. Through these three basic components, students are introduced to balancing competing time demands for themselves and others. Successful students learn to manage their time well in meeting their academic and non-academic responsibilities at the Academy. In order to meet the requirements of graduating students in 4 years, there is significant mentoring and advising that takes place throughout students' years at the Academy. Three advisors (academic, military, and athletic) work with each student to monitor their intellectual development and personal and professional growth. As previously mentioned, the competencies that students develop through this process are grouped into four categories and listed in Table 3.1. These categories form the framework and are discussed in the context of the three-basic set of activities shown in Figure 3.2 in the following sections.

Participation in non-academic activities outside of the classroom is part of the USCGA culture; students make the time to fit these extracurricular activities into their busy schedule. Students are also encouraged to participate in recognized social clubs. USCGA is unique because the sense of community service is introduced during freshman year and reinforced yearly through a mandatory 8 hours of community service per semester and other military training. Although students have the freedom to select community programs of interest, the overall impact of these activities on the outlook and preparedness of civil engineering graduates has been very positive. Community service is also organized through the student professional engineering chapters such as the American Society of Civil Engineers (ASCE) and Society of American Military Engineers (SAME). ASCE student members have participated in several Habitat for Humanity Construction projects within the local community.

All students are involved in athletic or intramural sports and encouraged to participate in at least two seasons of sports each year. In addition, students coordinate and run the intramural sports program and serve in leadership roles within their chosen sport or activity. Throughout these activities, students practice teamwork, goal setting, diversity appreciation, tolerance, conflict management, and communication. Student life at USCGA is fast paced, and there are limited opportunities to repeat courses or extend the length of study. All students are expected to manage their time effectively to meet the academic and non-academic rigors. They begin to see the "big picture," develop a sense of purpose, serve others, and be motivated to succeed. Students develop effective study habits, time management, and leadership skills that serve them well during their career in the Coast Guard and Civil Engineering.

While the curriculum at USCGA provides foundational knowledge and skills to graduate civil engineers who are ready to lead and manage effectively, students are given frequent opportunities to put these skills into practice outside of the classroom through their participation in military, athletic, and other extracurricular activities. Students have opportunities to take active leadership roles in the running of daily life on campus and managing other underclass cadets. For example, sophomore students serve as mentors to the freshman, overseeing the new students' transition to military, college, and academy life. Juniors and seniors, in addition to leading the freshman and sophomores, play an active leadership role in running the day-to-day operations of the Academy and the Corps of Cadets. Additional examples such as various groups of students participate in planning the annual Ethics Forum, Parent's Weekend, Homecoming, Research Symposium, and Graduation ceremony. The Undergraduate Research Symposium is a day set aside in the spring semester in which students showcase the results of their capstone projects. During the symposium, students present their senior capstone design project to a diverse audience consisting of faculty, staff, industry leaders, Coast Guard officers, and other stakeholders. There are several forums through which cadets are encouraged to propose ways of improving how things are done at USCGA, hence affecting change and improving performance where applicable. In general, cadets have a voice in making some of the decisions that affect life on campus. By graduation, each cadet would have had several opportunities to serve in positions that involve leading other cadets thus developing and honing the competencies in these categories listed in Table 3.1.

USCGA civil engineering students are actively involved in academic clubs such as ASCE, Society of American Military Engineers (SAME), and the Moles (an organization of individuals engaged in heavy construction). While not all colleges have a Corps of Cadets and an all residential student body, engineering students at non-military institutions should be encouraged to take on leadership and mentorship roles. In most academic institutions, involvement in student governing organization is typically dominated by non-STEM (science, technology, engineering, and mathematics) students. Engineering students are the busiest with little or no time to explore other avenues for their leadership development. However, there are opportunities to participate in ASCE student chapters, engineering honor societies, Habitat for Humanity, Engineers Without Borders, drama clubs, musical activities, local community volunteer activities, and partnership with local schools, other sports and cultural clubs on campus, or the local community. Academic institutions should encourage students to participate in non-academic activities outside the classroom that serve as avenues for developing and practicing leadership skills.

3.4 CLOSING THOUGHTS

Today's engineering programs must educate, train, and prepare future leaders so that they will be capable to operate and lead their organizations in a dynamic environment that is increasingly more global, complex, and unpredictable. An effective leadership development model is one that relies on integrating theory of leadership with practice and experiential learning through extracurricular and service-learning experiences that are supported by academic and non-academic or community-based activities. A distinctive component of this approach comes from the specific leadership competencies that are required to be developed in order to produce effective leaders. Those competencies could be grouped into categories that enable them to be infused into an academic program at various stages in the curriculum.

The strategy of leadership development practiced at USCGA represents a framework that can be adopted by other academic undergraduate programs to ensure the development of effective and ethical leaders. The leadership model used at USCGA includes 28 competencies divided into four categories (leading self, leading others, leading performance and change, and leading the Coast Guard). The development of these competencies is incorporated into the curriculum and reinforced both in and out of the classroom through academic, non-academic, and professional practice activities. This model has been successfully used to educate, train, and develop leaders of character who are ethically, intellectually, and professionally prepared to serve as civil engineers and officers in the Coast Guard. Civil engineering graduates serve through active duty assignments as well as in leadership positions in the public and private sectors after leaving the Coast Guard. Aspects of this model could be adopted by other academic institutions as they strive toward preparing future civil engineering leaders for the profession.

REFERENCES

ABET (2018), *Criteria for Accrediting Engineering Programs*, ABET Engineering Accreditation Commission, Baltimore, MD.

American Society of Civil Engineers (2019), *Civil Engineering Body of Knowledge, Preparing the Future Civil Engineer*, 3rd Edition, ASCE Press, Reston, VA.

Arendt, S.W. and Gregoire, M.B. (2005). "Leadership behaviors in hospitality management students." *Journal of Hospitality & Tourism Education*, Vol. 17, pp. 20–27.

Athreya, K.S., Kalkhoff, M., McGrath, G., Bragg, A., Joines, A., Rover, D. and Mickelson, S. (2008). Work in progress – engineering leadership program: tracking leadership development of students using personalized portfolios. *38th ASEE/IEEE Frontiers in Education Conference*, Saratoga Springs, NY, October 2008.

Bloom, B.S. (Ed.). Engelhart, M.D., Furst, E.J., Hill, W.H. and Krathwohl, D.R. (1956). *Taxonomy of Educational Objectives, Handbook I: The Cognitive Domain*. David McKay Co Inc, New York.

Brown, L. and Posner, B. (2001). "Exploring the relationship between learning and leadership." *Leadership and Organization Development Journal*, Vol. 22, pp. 274–280.

Commandant Instruction COMDTINST M5351.3. (2006), *Leadership Development Framework*. United States Coast Guard, Washington, DC.

Day, D.V. (2001). "Leadership development: a review in context." *Leadership Quarterly*, Vol. 11, No. 4, pp. 581–613.

DeLisle, P.A. (1999). "Engineering Leadership," The Balanced Engineer: Entering a New Millennium, *Proceedings of IEEE-USA Professional Development Conference*, Dallas, Texas.

Dodgson, M. (2011). "Exploring new combinations in innovation and entrepreneurship: social networks, Schumpeter, and the case of Josiah Wedgwood (1730–1795)." *Industrial & Corporate Change*, Vol. 20, No. 4, pp. 1119–1151.

Hawken, P. (2007). *Blessed Unrest: How the Largest Movement in the World Came into Being and Why No One Saw it Coming*. Penguin Group, New York, NY.

Houghton, J.D. and Neck, C.P. (2002). "The Revised self-leadership questionnaire: testing a hierarchical factor structure for self-leadership." *Journal of Managerial Psychology*, Vol. 17, pp. 672–691.

Kaiser, R.B., Hogan, R. and Craig, S.B. (2008). "Leadership and the fate of organizations." *American Psychologist*, Vol. 63, pp. 96–110.

Kouzes, J.M. and Posner, B.Z. (2011). "Leadership begins with an inner journey." *Leader To Leader*, Vol. 2011, No. 60, pp. 22–27.

Kouzes, M., Posner, J. and Biech, E. (2010). *A Coach's Guide to Developing Exemplary Leaders: Making the Most of the Leadership Challenge and the Leadership Practices*. Inventory (LPI). Jossey–Bass, San Francisco, CA.

Maellaro, R. (2013). The learning journal bridge: from classroom concepts to leadership practices. *Journal of Leadership Education*, Vol. 12, No. 1, pp. 234–244.

Manz, C.C. and Sims, H.P., Jr. (2001). *The New Super-Leadership: Leading Others to Lead Themselves*. Berrett-Koehler, San Francisco, CA.

National Academy of Engineering. (2004). *The Engineer of 2020: Visions of Engineering in the New Century*. National Academy Press, Washington, DC.

Neck, C.P. (1996). "Thought self-leadership: a self-regulatory approach to overcoming resistance to organizational change." *International Journal of Organizational Analysis*, Vol. 4, pp. 202–216.

Northouse, P.G. (2007). *Leadership: Theory and Practice*. 4th Edition, Sage Publications, London.

Prussia, G.E., Anderson, J.S. and Manz, C.C. (1998). "Self-leadership and performance outcomes: the mediating influence of self-efficacy." *Journal of Organizational Behavior*, Vol. 19, pp. 523–538.

Riggio, R.E., Riggio, H.R., Salinas, C. and Cole, E.J. (2003). "The role of social and emotional communication skills in leader emergence and effectiveness." *Group Dynamics: Theory, Research, and Practice*, Vol. 7, No. 2, pp. 83–103.

Skipper, C.O. and Bell, L.C. (2006). "Influences impacting leadership development." *ASCE Journal of Management in Engineering*, Vol. 22, No. 2, pp. 68–74.

Stogdill, R.M. and Bass, B.M. (1981). *Stogdill's Handbook of Leadership: A Survey of Theory and Research*. The Free Press, New York.

Tarabishy, A., Solomon, G., Fernald Jr, L.W. and Sashkin, M. (2005). "The entrepreneurial leader's impact on the organization's performance in dynamic markets." *Journal of Private Equity*, Vol. 8, No. 4, pp. 20–29.

Taylor, R.L. and Rosenbach, W.E. (2005). *In Pursuit of Excellence*, 5th Ed. Westview Press, Boulder, CO.

Thayer, L. (1988). Leadership/communication: a critical review and a modest proposal. In G.M. Goldhaber, G.M. and Barnett, G.A. (Eds.), *Handbook of Organizational Communication* (pp. 231–263). Praeger, Norwood, NJ.

Todd, M.J. (1996, July/August). "21st century Praeger leadership and technology." *Journal of Management in Engineering*, Vol. 12, No. 4, pp. 40–49.

Wheatley, M. and Frieze, D. (2011). *Communities Daring to Live the Future Now*. Berrett-Koehler Publishers, Oakland, CA.

4 Communication and Information Literacy Skills

4.1 INTRODUCTION

It has been reported in the literature that effective verbal/oral and written communication skills are one of the primary factors required of new engineering graduates that ultimately affect their success in the workforce. Riley et al. (2000) reported the results of a study showing 38% of new engineering graduates across all engineering specialties indicated that effective communication skills are important factors affecting their advancement and success in industry. They also reported that effective communication is the single area where they perceive the greatest gap exists with respect to their engineering preparation in college. Professional engineering organizations and universities continually list the ability to communicate technical information as a highly sought attribute in new graduates. One of the ABET Student Outcomes, *Ability to Communicate Effectively with a Range of Audience*, has required students to develop communication skills within the engineering curriculum for the last two decades. This requirement is supported by the American Society of Civil Engineers third edition of the *Civil Engineering Body of Knowledge* outcome on Communications (ASCE BOK3, 2019). Practicing engineers are expected to not only prepare technically appropriate designs but to communicate these designs in written, oral, and graphical forms to a variety of audience ranging from their technical peers to the general public. Engineers who are good public speakers can raise their visibility within a company and will probably be asked to lead project teams, present their work in front of senior management, lead site tours, and represent the organization at important forums and meetings. ASCE BOK3 expects civil engineering graduates to be acquainted with the tools used to communicate their work using various forms of communications techniques. Civil engineers are expected to communicate effectively and persuasively to technical and non-technical audiences in a variety of settings using formal and informal means. Therefore, it is essential that engineering educators encourage and guide students to develop and improve their communication skills in the context of the current and emerging information infrastructure.

It is important for students to know that they will need to be able to communicate to different audiences and that it is crucial for them to learn how to achieve the appropriate tone and style for the intended audience. The inclusion of writing in technical courses stresses the importance of writing in developing engineers and encourages them to develop the necessary communication proficiency desired in the profession. Also, proficiency in technical writing takes time and must be developed progressively, through consistent practice and efforts over the 4 years of undergraduate engineering education but not over a single semester. Ford and Riley (2003) highlighted examples

DOI: 10.1201/9781003280057-4

from 17 universities across North America to offer useful portraits of writing across the curriculum approaches, interdisciplinary courses, integrated programs, and a variety of support systems including writing and communication centers and online resources at those universities.

With the current trends in texting and instant communications, online access provides tremendous amounts of information (true and fake) that could be quickly downloaded and used. As a result, Information Literacy (IL) has become an important element of undergraduate education in response to these growing number of massive information sources. Niedbala and Fogleman (2010) argued that developing IL competency is critical for students' continued professional career and life-long learning. Engineering educators must ensure that students learn the important steps of evaluating the credibility and appropriateness of sources while properly synthesizing, using, and citing information from several sources. The National Forum on Information Literacy and the American Library Association (1989) define IL as *The ability to know when there is a need for information, to be able to identify, locate, evaluate, and effectively and responsibly use and share that information for the problem at hand.* Moll (2009) advocated the integration and development of IL skills across curricula and recommended application of these skills in real-life situations across various assignments, research papers, and presentations. Specifically, customized IL programs at any college-level benefit students by moving them from basic IL skills to a higher level of IL confidence, fluency, and proficiency. Grassian and Kaplowitz (2001) reported about initiatives that can be instilled throughout the curriculum and could morph into IL educational instruction that ranges from an undergraduate to a graduate program experience.

This chapter provides an overview of communication and IL skills development strategies. A case study of the approach at USCGA to embed and evaluate aspects of communication effectiveness and IL within the Civil Engineering curriculum without adding new courses is presented. The progressive infusion of IL and communication skills into the curriculum, the development of relevant performance indicators, and steps taken to collect and analyze assessment data are also discussed.

4.2 EXAMPLES OF COMMUNICATION SKILLS DEVELOPMENT STRATEGIES

Writing has been shown to enhance active learning and critical thinking and addresses the needs of students with different learning styles. Writing can be used to assess student understanding when demonstrated as a process. Similarities between the writing and design processes can be used to highlight the fact that there is often no single "correct" solution to either and that feedback and revision are often crucial to both. In ASCE BOK3, the framework for achieving the appropriate level of communication skills is summarized in Table 4.1. The ASCE BOK3 Communication Outcome is expected to be fulfilled by undergraduate education only up to the third level of Bloom's Taxonomy. Higher levels of Bloom's cognitive domains (analyze, synthesize, and evaluate) are expected to be addressed in industry through practical

TABLE 4.1
Civil Engineering Body of Knowledge "Communications" (ASCE BOK3, 2019)

Cognitive Domain Level of Achievement	Demonstrated Ability	Fulfilled Through
1 – Remember (remember previously learned material)	**Identify** concepts and principles of effective communications for technical and non-technical audiences.	Undergraduate Education
2 – Comprehend (grasp the meaning of learned material)	**Explain** concepts and principles of effective communications for technical and non-technical audiences.	Undergraduate Education
3 – Apply (use learned material in new and concrete situations)	**Formulate** effective communications for technical and non-technical audiences.	Undergraduate Education
4 – Analyze (break down learned material into its component parts so that its organizational structure may be understood)	**Analyze** effective communications for technical and non-technical audiences.	Mentored Experience
5 – Synthesize (put learned material together to form a new whole)	**Integrate** different forms of effective communication for technical and non-technical audiences.	Mentored Experience
6 – Evaluate (judge the value of learned material for a given purpose)	**Assess** the effectiveness of communication for technical and non-technical audiences.	

experiences and mentorship. Details of Bloom's Taxonomy including the levels were discussed in Chapter 2.

Plumb and Scott (2002) developed and implemented the Engineering Writing Assessment Program at the University of Washington, Seattle, Washington. Engineering students were required to take a course *Introduction to Technical Writing* and a recommended elective course *Advanced Technical Writing and Oral Presentation*. The introductory technical writing course presented basic concepts of technical writing, such as writing for targeted audiences, and provided practice in genres that students will encounter in school and work, such as research reports, instructions, and newsletter articles on technical subjects. The advanced course provided instruction and practice in oral presentation as well as the opportunity for more writing practice, primarily in workplace genres such as resumes, progress reports, and recommendation reports. The assessment program used by Plumb and Scott consisted of two processes: (1) entry and exit surveys of students to assess competence in the broad concepts and attitudes listed in the outcomes; and (2) holistic and fine-grained review of student papers to assess competence in the specific writing skills listed in the outcomes. The assessment results revealed weaknesses that span a broad range of writing issues and included all the four areas they focused on in the technical writing instruction: content, organization, design, and mechanics/style. The improvement loop was to identify that students needed more practice, more individual help, more feedback with their writing, and more instructions on using and citing sources. Assessment results also revealed that instructors needed to increase emphasis on the importance of strong introduction, including stating the purpose and defining the scope. The assessment data showed that weak content may indicate a lack of topic knowledge, and poorly substantiated claims may indicate muddled thinking. These writing assignments challenged students to research their topics and to provide evidence of that research. These model assignments gave students clear instructions, described the audience for the assignments, and specified how the assignments would be evaluated.

Daniell et al. (2003) reported a case study about their experience in improving teaching technical writing within Mechanical Engineering undergraduate laboratories at Clemson University, South Carolina. Mechanical Engineering faculty require students to "write to learn" technical concepts, applications, and problem-solving strategies by collaborating with English faculty to design assignments in which students "learn to write." Several active-learning and communication projects were developed and implemented in some engineering courses. So, it was natural to collaborate again on developing or at least improving technical writing skills through the framework of the engineering laboratories. Engineering faculty were expected to prepare students to become proficient in writing engineering lab reports by providing instructions that include: (1) developing audience awareness and using engineering language appropriate to that audience; (2) understanding types, or genres, of writing, such as the executive summary, lab report, technical paper; (3) developing effective argument through analysis and principles and with a clear presentation; (4) clearly expressing logical thinking and deductive reasoning; and (5) defining a question and developing an answer to that question. Using end of semester survey as an assessment tool, students offered suggestions for improved learning, including the request for more feedback on report section examples and writing style. Other aspects students conveyed as effective included the opportunity to revise and resubmit with good feedback and an opportunity for generous grade improvement, good sample papers to use as models, emphasis on audience, use of the passive voice, and peer review. English faculty served as consultants to review student work, instructor feedback, and interactions. Writing style and communication are tied closely with genre, audience, and context. The consultants offered several specific suggestions to enhance the writing experience, including rewriting with constructive feedback, use of scenarios to define context, and peer review.

Conrad et al. (2012) reported the results of a project that investigated the characteristics of effective writing in civil engineering practice and improved writing instruction for students. The project demonstrated the different viewpoint of writing: practitioners see writing as integrated with engineering content and practice, whereas students view writing as separate from engineering. Becoming proficient with workplace writing requires years of practice in the workplace context. Writing effectively requires skills and judgment, and these skills and judgment take just as long to develop as other engineering skills. The project offered five specific teaching suggestions for approaching writing in civil engineering classes so that students will be better prepared for writing in the workplace.

1. Establish the importance and amount of writing in civil engineering practice from the beginning of the program and reinforce it at each level. Review the types of writing civil engineers produce, discuss the importance of writing for advancing their careers, and review basic principles for making written explanations precise and accurate.

2. Explain writing and give feedback on writing as connected to meaning and the needs of clients, not just to formatting and style. Address the choice of active and passive voice, and the use of active voice and personal pronouns for speaking directly to clients or establishing a firm's responsibility clearly.

3. Provide students with a target for organizing their papers. In practice, most documents will have a specified format, and clients will typically either expect or demand that certain information is covered in certain sections. Therefore, faculty will need to explain and exemplify what the target is when students face a new assignment type.

4. Show students examples of effective writing and explain what makes it effective. Even when students are told writing principles to follow or organization to use, they often have difficulty knowing exactly how to manipulate the language on their own papers or reports.

5. Reinforce the use of standard written English and effective proofreading as an important way to establish credibility as a professional. Preparing students for workplace practice means demonstrating to students that knowing Standard English grammar is part of engineering practice. Students need consistent feedback telling them that obviously inaccurate, ambiguous statements or numerous grammatical errors are not acceptable.

Harichandran et al. (2014) reported on the implementation of a Project to Integrate Technical Communication Habits (PITCH) across seven engineering undergraduate programs and computer science at the University of New Haven, New Haven, Connecticut. The goal of PITCH was to emphasize professional communication skills and professional habits across engineering disciplines. A comprehensive set of learning outcomes were developed based on surveys of engineering faculty, alumni, and employers. Communication assignments were developed based on engineering content and designed to have students achieve stated outcomes in a developmental progression throughout their undergraduate programs. The project leveraged technology to provide students and faculty with supporting resources. To ensure that the PITCH outcomes would be met at the time of graduation, technical communication products or simulation and specific technical communication habits were distributed among course sequences in each of the seven engineering programs. These distributions were planned to introduce skills and habits in introductory courses and reinforced in advanced courses reaching deeper engineering content and more complex communication situations. Technical communication such as letters, technical memoranda, short reports, and formal e-mails were implemented in four courses that are part of this curriculum. Reports documenting experimental or simulation methods and results were implemented in sophomore and junior year disciplinary courses, and formal reports (proposals, analyses, progress reports, and design documents) were implemented in senior design courses. Another technical communication product is to plan, prepare, and deliver oral presentations along with poster displays in capstone. In addition, eight communication skills were emphasized throughout the 4 years: (1) use appropriate format and content; (2) exhibit clear, precise, and logical expression; (3) demonstrate appropriate organization, level of detail, style, and tone for a given audience, situation, and purpose; (4) demonstrate appropriate syntax and correct usage of grammar and spelling; (5) highlight or identify critical information; (6) present, discuss, and summarize data accurately and persuasively; (7) write thoughtful and persuasive conclusions and recommendations; and (8) work effectively to produce multi-author communications. PITCH was designed to help students develop written, oral, and visual communication skills starting in the first semester and continuing through all 4 years of each program. Specific courses that span all 4 years are targeted for implementation and assessment of technical communication skills. The different communication instruments are appropriately distributed across courses, and the skills are developed at deeper levels as students' progress through the years. A critical feature of the project is that technical communication skills are integrated into the content of regular engineering courses and are taught by regular engineering faculty.

Conrad (2017) reported the results of another study that compared the word-level, sentence-level, and organizational difference in writing by practitioners and students and to identify differences that are important to the engineering practice. The study investigated four analyses covering organization, grammar features, and word choices in practitioner and student writing in civil engineering. The analyses were chosen to illustrate the range of features that can be addressed with linguistic techniques. The study demonstrated that student writing had more complicated sentence structures, less accurate word choice, more errors in grammar and punctuation, and less linear organization. These characteristics decreased effectiveness in areas that practitioners considered important such as accurate and unambiguous content, fast and predictable reading, and attention to details. Underlying the student writing problems were misconceptions about effective writing, ignorance of genre expectations, weak language skills, and failure to appreciate that written words, not just calculations, express engineering content. The results of this study defined the gap between student and practitioner writing and are the basis for instructional materials that target important student writing weaknesses.

Furthermore, it is known that social media plays a role in student writing habits and skills, where abbreviated words and "emoticons" take the place of sentences and complete thoughts. With the current trends in texting and instant communications, online access provides tremendous amounts of information that could be quickly downloaded and used by students. In addition, automated spelling and grammar correction and the endless amount of information on the Internet seemingly remove the need for attention to writing and research skills. In many cases, students are no longer required to participate in tasks such as data gathering, creating an outline, or selecting references, because a simple Google search will often yield a ready-made template for their writing assignments. For these reasons and others, many students do not devote the time, energy, or thinking required for creating quality technical documents. As a result, engineering educators should ensure that students develop their communication skills and learn the important steps of evaluating the credibility and appropriateness of sources while properly synthesizing, using, and citing information from several sources. To improve student development in technical writing, faculty members must develop assignments and grading rubrics designed to consistently assess student progress during their undergraduate experience. Therefore, technical writing and IL should be progressively and consistently incorporated into the undergraduate engineering curricula for students to develop and hone the required skills necessary for engineering practice and life-long learning.

4.3   IL DEVELOPMENT

In general, one of the main objectives of higher education is to prepare students to be life-long learners, critical thinkers, and to take greater responsibility for their own learning. This cannot be accomplished without the appropriate competency in IL. The Association of College and Research Libraries (ACRL2000) states that:

> Information Literacy forms the basis for life-long learning. It is common to all disciplines, and to all levels of education. It enables learners to master content and extend their investigations to become more self-directed and assume greater control over their own learning.

This is in harmony with ABET requirements on communications and life-long learning for the accreditation of engineering programs. It is a critical thinking process that is iterative and linked to the acquisition and practice of discipline knowledge. Engineering educators must ensure that students learn the important steps of evaluating the credibility and appropriateness of sources while properly synthesizing, using, and citing information from several sources.

Doyle (1992) reported that the concept of IL originated in 1914. Since that time, the term has gone through numerous changes in its definition, application, and assessment. As computer technology advanced, information available to students increased and access improved; it became vital to tie fundamental concepts of learning skills across all fields to the development of IL skills. Saleh (2013) noted that IL is becoming increasingly important in the contemporary environment of rapid technological change and proliferating information. Cox and Lindsay (2008) argued that IL supports development of critical thinking skills, reflection skills, and independent learning skills in the context of the increasingly extensive amounts of information that is available through a wide range of technologies, sources, and modes. These skills are required as they enable students to access and navigate the growing spheres of information, to appropriately select credible and reliable information, to read critically and think independently as they create their own ideas, and to use that refined information in a variety of venues. Swanson (2011) stresses that instructors and librarians should work together to create critical IL models that:

- Views the information world as a dynamic place where authors create knowledge for many reasons.
- Seeks to understand students as information users.
- Emphasizes that information evaluation is a continual process during research.

- Recognizes that information evaluation is relative to the point of view of the reader.
- Provides opportunities for students to increase their understanding of finding, evaluating, and using information.
- Centers libraries within the curriculum as the experts on overcoming many of the obstacles to conducting successful research in the ever-changing information world.

According to the literature on IL, some of the attributes of an information literate individual include the ability to: (1) obtain and use information, (2) make use of available technologies, (3) acquire and use all other available resources effectively to find and manage information, and (4) critically evaluate and ethically apply that information to solve a problem and conduct basic research analysis. Other characteristics of an information literate individual include the spirit of inquiry and perseverance to find out what is necessary to get the job done. Some authors define IL as a skill or learning tool, others refer to information attitude or study and research skills or how to think critically (Bruce, 1997; Campbell, 2004; Owusu-Ansah, 2003), but all definitions of IL contain some common elements. Each definition mentions possession of an integrated set of skills, knowledge of resources from which to retrieve information, and the ability to analyze and use information (Burkhardt et al., 2003; Rockman, 2004). A standard definition of IL that is widely used was established by the ACRL (2000) and it defines IL as: "…a set of abilities requiring individuals to recognize when information is needed" and an "ability to locate, evaluate, and use effectively the needed information" (Source: http://www.ala.org). The National Forum on Information Literacy (IL) defines IL as "The ability to know when there is a need for information, to be able to identify, locate, evaluate, and effectively and responsibly use and share that information for the problem at hand." Therefore, key steps in practicing IL could be summarized as:

- Identify the need for information
- Identify and locate sources of information
- Evaluate the credibility of these sources
- Responsibly use and share the acquired information

Some of the most relevant and important aspects of IL are summarized as follows:

- IL is seen as the domain of librarians, but research shows that collaboration between librarians and faculty is crucial for successfully producing information literate students. IL principles become most meaningful in the context of a discipline, so librarians and instructors from all disciplines must work together (Jarson, 2010).
- Integrating IL principles into the research and writing process empowers students by teaching them expert-level skills in their information use. By nature it makes writing a deliberate and process-based endeavor. Students often report that IL-based exercises make them feel confident, creative, and proficient (Ratteray, 1985 and 2000–2002).

- Assessment of student learning is most meaningful when it is reliable and consistent. Assessment with rubrics has been shown to be a viable and effective option for authentically assessing IL-based student work. Rubric assessment of writing samples has shown that this approach results in better organization of thoughts and information use, selection and interpretation, and consequently better writing overall across all disciplines.
- IL competencies at the undergraduate level are required by several accreditation agencies such as ABET. These accreditation requirements ensure that academic programs provide students with the tools needed to develop IL skills.

In practice, IL can be taught either as an independent course or integrated into several courses within an academic major. However, it is more effective when IL is implemented and used across all courses from freshman through senior year. To facilitate IL development at academic institutions, five key aspects should be considered: (1) IL Competencies, (2) Current Library Instruction, (3) IL across Curriculum, (4) Evidence of Students Learning, and (5) Resources for Assessment and Evaluation. These components are briefly discussed as follows:

1. *IL Competencies* – To determine the appropriate IL competencies, a working definition should first be adopted. The most common definition is that from the National Forum on Information Literacy that defines IL as "the ability to know when there is a need for information, to be able to identify, locate, evaluate and effectively use that information for the issue or problem at hand." The Association of College and Research Libraries (ACRL) has developed seven competences that are widely used. These include the ability to:
   1. determine the extent of information needed
   2. access the needed information effectively and efficiently
   3. evaluate information and its sources critically
   4. incorporate selected information into one's knowledge base
   5. use information effectively to accomplish a specific purpose
   6. understand the economic, legal, and social issues surrounding the use of information
   7. access and use information ethically and legally (the ACRL Information Literacy Competency Standards for Higher Education, 2000).
2. *Library Instruction* – Librarians have the primary responsibilities of facilitating access to information resources and providing instruction on how to use those resources. As such, access to the appropriate library resources plays an important role in helping students develop and enhance their IL skills. Therefore, librarians should be active partners with faculty in the effort to promote IL and help students develop those skills.
3. IL *across the Curriculum* – To be effective, IL should be continuously and progressively infused into the entire curriculum over the 4 years of undergraduate education. It is least effective when only addressed in a single course. Specific components must be infused across years and the importance to life-long learning emphasized.

4. *Evidence of Student Learning* – As IL augments students' ability to evaluate, manage, and use information, expand knowledge, effective communication, and critical thinking ability, students should be regularly assessed in these areas. In addition to discipline-specific content, individual course contents should be designed to measure one or more selected competencies that are appropriate for the specific course. IL competency extends learning beyond formal classroom settings and provides practice with self-directed investigations as students move into specific research projects, internships, and professional positions after graduation.

5. *Resources for Assessment and Evaluation* – A well-developed and comprehensive IL program assessment plan is essential to improve outcomes in academic IL initiatives. In order to target instruction effectively, to offer needed support, and to provide appropriately challenging assignments, an ongoing and accurate understanding of students' IL capabilities is required. The resources needed to support an IL program include personnel, fiscal responsibility, technology, and other services. Faculty and librarians should provide IL instruction and continue developing a curriculum in an atmosphere of collaboration, including professional development opportunities and incentives. IL strategic approach requires providing funding for adequate resources and creating opportunities for collaboration and staff development among faculty, librarians, and other professionals. The challenge in providing an effective IL program is to continuously assess comprehensive range of strategies including resources and collaboration between faculty and library staff.

4.4   CASE STUDY: COMMUNICATION AND IL SKILLS DEVELOPMENT AT USCGA

The USCGA is an undergraduate academic institution that focuses on educating, training, and developing leaders of character who will serve as leaders in the USCG. To accomplish this, the USCGA developed and uses five basic shared learning outcomes that promote meeting the mission objectives of the academic and professional training programs. These USCGA learning outcomes are presented and explained in the following paragraphs:

- *Leadership Abilities* – Graduates shall be military and civilian leaders of character who understand and apply sound leadership principles and competencies. This includes the ability to direct, develop, and evaluate diverse groups; to function effectively and ethically as a leader, follower, facilitator, or member of a team; and to conduct constructive assessment of self and others.

- *Personal and Professional Qualities* – Graduates shall maintain a professional lifestyle that embraces the Coast Guard Core Values of *Honor, Respect, and Devotion to Duty*, includes physical fitness and wellness, and demonstrates the customs, courtesies, and social skills befitting members of a maritime military service. Graduates shall also have a sense of Coast

Guard maritime heritage and an understanding of the roles that the Coast Guard and the nation play in the global environment.

- *Communication Effectiveness* – Graduates shall be able to write clearly, concisely, persuasively, and grammatically; prepare and deliver well-organized and polished oral presentations; read and understand a variety of written materials; listen thoughtfully to oral arguments; respect diverse opinions; and formulate reasoned alternatives and responses.
- *Ability to Acquire, Integrate and Expand Knowledge* – Graduates shall have developed the motivations and skills for "lifelong learning." Graduates shall be able to create a working conceptual framework that lends itself to continued expansion. To accomplish this, graduates shall be able to efficiently access a broad range of information sources, locate and interpret desired data reliably, employ appropriate technology, and integrate knowledge. Graduating cadets shall also have acquired and integrated the specific in-depth knowledge required of both an academic major and an entry-level professional assignment.
- *Critical Thinking Ability* – Graduates shall be able to accomplish complex tasks in a broad range of contexts by applying the basic skills of critical analysis, systems thinking, quantitative reasoning, risk management, creative problem solving, and value-based decision-making.

These shared learning outcomes are part of the core curriculum and the Civil Engineering Program. Communications and Information Literary instructions are introduced in the introductory English courses and integrated into the required civil engineering courses that are taught by regular engineering faculty. During 4 years of undergraduate studies, students are required to develop adequate proficiencies in these outcomes. The Hewitt Writing and Reading Center was established to consistently aid faculty and students in the institutional efforts to improve communications and IL skills. The Writing Center is staffed by faculty members from the various academic disciplines, professional editors, writers, and educators and provides support to: (1) faculty as they prepare reports or articles for academic and professional journals and (2) students as they prepare laboratory reports, technical papers, writing assignments, and presentations in their courses. The partnership between the Library, Writing Center, and faculty at USCGA is an effective collaboration for developing the student shared learning outcomes to communicate effectively and engage in life-long learning.

The Civil Engineering Program developed specific performance indicators related to communications that have been linked to several of the ABET Student Outcomes (i.e. ABET "1–7"). Faculty members have developed assignments and rubrics designed to assess student progress and improve student development in technical writing and IL for each performance indicator. Faculty are constantly challenging students to use their communication skills progressively throughout the curriculum in order to achieve the highest cognitive level in Bloom's Taxonomy. With higher levels of communication skills, students will significantly improve their ability to communicate effectively with a wide range of audience. By integrating writing, IL development, and assessment into the existing civil engineering assessment model, the faculty has successfully threaded these competencies into the curriculum using a sustainable and effective framework. Part of the framework involves using similar rubrics throughout the curriculum that includes assessing student's ability to: (1) identify the type and extent of information needed; (2) search for and incorporate a variety of appropriate technical information sources; (3) use sources appropriately, legally, and ethically; (4) use technical writing skills; and (5) write a well-organized paper or report. Details of the assessment are discussed in Chapter 6.

4.4.1 WRITING ACROSS THE CIVIL ENGINEERING CURRICULUM

Writing across the curricula at USCGA starts by introducing first-year students to common language which comprises of 12 rhetorical terms that were distilled from a typical first-year composition handbook (Jernquist & Clippinger, 2017). The 12 rhetorical terms are: audience, purpose, thesis, voice, tone, stance, organization, development, style, diction, editing and conventions. The basic chart of rhetorical language for thinking about communication effectiveness is presented in Table 4.2. This chart is made available to faculty throughout the Academy for adoption into their courses. The terms are structured in three categories designed to reflect the learning and writing processes through which students typically work.

The rhetorical terms are applied in first-year writing, in preparing faculty to mentor students at the Writing Center, and in supporting STEM faculty in teaching technical writing in their courses. The common rhetorical language supports faculty in stating expectations for student writing. With assignments written in a language spoken among faculty and Writing Center personnel, writers can understand the language through repeated exposure. The Writing Center provides one-on-one learning by using language shared across departments that can contribute to campus-wide conversation about writing. The common language guides students to "identify points of contrast and common ground between disciplines and subfields." It also supports "the transfer of knowledge and skills from one course to another" and creates a coherent writing curriculum. Such a curriculum would support faculty in identifying "domains of writing knowledge that comprise writing expertise": content, process, rhetorical, genre (type of assignment), and discourse community (audience). Embedding the common language in assignments and assessments can support engineering faculty in teaching their students to communicate effectively. The common language was created for the teaching and learning of writing in core and major courses at USCGA.

The overall rhetorical language used in the writing process is summarized in Table 4.3. It shows the complexity of the writing process that includes simultaneously learning new material and genres for communicating that knowledge with effective sentences and disciplinary conventions. The 12 common language terms are listed in the first column of Table 4.3. These terms are used and applied to the early stages of writing (second column) where the audience is initially the writer in the process of learning the material. The writer starts with organizing and developing that knowledge in appropriate sections with the appropriate purpose of informing, analyzing, and persuading to meet audience expectations. The third column in Table 4.3 suggests expectations of the 12 rhetorical terms for writing the final draft. The use of this

TABLE 4.2
USCGA Common Rhetorical Language for Thinking about Communications (Jernquist & Clippinger, 2017)

Higher Order Concepts at the Essay Level	Structure and Coherence at the Paragraph Level	Structure and Coherence at the Sentence Level
Appeal to Audience • The instructor • Community peers • Community in particular discipline • Academic or military community	*Effective Strategies for Organization* • Introduction • Body paragraphs with transitions • Conclusions	*Effective Style* • Sentence length, rhythm, and clarity • Appropriate syntax • Concise expression
Achievement of Purpose • Information • Expression • Argument or persuasion • Literacy	*Effective Strategies for Development* • Definition • Explanation and examples • Analysis of cause and effects • Analysis of a process • Comparison and contrast • Classification and division • Narration	*Effective Diction* • Precise word choice • Control over language of the topic and discipline
Expression of Thesis and Topic • Focused statement or paragraph • Stated or unstated depending on audience, purpose, and topic		*Effective Grammar, Punctuation, and Mechanics*
Appropriateness of Voice, Tone, and Stance		*Effective Adherence to Conventions* • MLA (Modern Language Association) or appropriate format • Attention to the details of the assignment

table at USCGA has guided faculty and students in grasping the intellectual, time demand of learning how to write in any major. The stages listed in this table demonstrate to engineering students the importance of following a structure, developing a draft of the paper or report, and seeking help early from the Writing Center.

One of the objectives was to infuse communications into the current curriculum without developing or adding new courses. To achieve this, a process of identifying communications related to ABET student outcomes, linking them to courses, and developing assessment tools was established. The first step was to review all the courses that had a writing and/or research component throughout the 4-year curriculum. Civil Engineering students typically take 27 credits of non-technical courses such as *College Composition, Cultural Perspectives, Global Studies, American*

TABLE 4.3

USCGA Sample Rhetorical Language for a Writer's Stages of the Writing Processes (Jernquist & Clippinger, 2017)

	Writing Notes and Drafts	Writing Final Paper
Audience	Is yourself as you take notes, outline, and learn about your topic and the assignment	Is a reader whom you show that you have control over the material and the elements of writing
Purpose	Is to inform yourself about the material and assignment; to begin to persuade or inform a reader	Is to inform your reader about your knowledge of the material or argue / persuade a point convincingly
Thesis	Is a "working thesis" that will start generally and get more specific as you read, draft, and revise	Is a clear, focused statement of your main point or argument and your method to prove your point
Voice	Is casual, tentative, speculative	Is authoritative
Tone	Is informal	Is formal; factual for informative papers; varies for the argument
Stance	Is close to the writer as reader	Is professional, distant from reader for informative; varies for argument
Organization	Follows the order of the assignment question or directions	Follows clear introduction; body paragraphs have topic sentences and transitions; conclusion is clear
Development	Follows the specific language of the assignment, e.g. compare, contrast, define, give examples	Each paragraph develops with evidence and relates to the thesis and assignment
Style	Is informal	Sentences are clear; for information transaction; for argument sentence length and rhythm vary to create a clear effect
Diction	Is informal	Shows control over the language of the subject and formal writing
Editing	All choices about paragraph organization and development and sentence construction relate to the assignment and to the audience, purpose, and "working thesis"	Every element of organization, development, style, diction, editing and conventions relates to audience, purpose, thesis, and the assignment
Conventions	Adequate enough to draft	MLA or other required format

Government, etc. Each of these courses has some writing and IL component. The formal writing education that students receive is limited to the first-year English courses focused on expository writing, composition, and literature. These English courses must be broad-based to serve as foundation courses for all majors at USCGA. While these courses provide exposure to the writing process, thesis development, sentence structure, and grammar, they do not necessarily help the student transition to the skills necessary for engineering technical writing. Since effective communication skills are as important to engineers as their technical skills, students need guidance to help them understand the complexities of the engineering writing processes and the expected outcome (report, memo, etc.). Students need to understand

and appreciate the investment of time required to develop effective communication skills, value audience expectations, and learn the forms to express content knowledge such as laboratory report or design project report. Students are expected to learn the conventions of sentence-level expression and develop a clear style that meets reader expectations for sentence length and rhythm along with graphs or tables that are properly labeled and formulas that are clearly explained. Although these contribute to students' overall competency in writing and IL, they do not adequately address technical writing relevant to Civil Engineering. Therefore, a decision was made to focus on courses directly controlled by the Civil Engineering Section that could be used to enhance technical writing skills in collaboration with the Writing and Reading Center at USCGA. Since the development of communication skills overlaps with the IL skills development, more details on how this is addressed are presented in the following subsection on IL.

4.4.2 IL ACROSS THE CIVIL ENGINEERING CURRICULUM

As a result of a program review and alumni feedback, the Civil Engineering faculty made efforts during the 2010 fall semester to specifically infuse components of IL into the curriculum and emphasize its importance to life-long learning. One of the objectives was to infuse IL into the curriculum without developing or adding new courses. To achieve this, Jackson et al. (2011) developed a process of identifying IL related to ABET student outcomes, linking them to courses, and developing assessment tools. The first step was to review all the courses that had a writing and/or research component throughout the 4-year curriculum. Specific performance indicators related to IL were developed and linked to several ABET student outcomes. The goal was to promote IL within the current civil engineering curriculum by developing assignments and assessing student achievement of performance indicators. Various components of IL were progressively infused throughout the curriculum to enable students to develop and foster the appropriate skills. Most recently, there has been an institutional effort to develop IL as a thread in the new core curriculum throughout USCGA. Faculty are actively promoting IL in close collaboration with the library staff so that life-long learning skills are valued and supported in the context of academic disciplines. A schematic representation of the process used to identify the IL components in the ABET student outcomes, link them to courses, and develop assessment tools is presented in Figure 4.1 (Jackson et al., 2011).

As mentioned earlier, Civil Engineering students typically take 27 credits of non-technical courses such as *College Composition*, *Cultural Perspectives*, *Global Studies*, *American Government*, *American Government*, etc. Each of these courses has some IL component. The courses in the CE curriculum that contribute to IL are presented in Table 4.4. Although courses outside of the CE department contribute to students' overall competency in IL, they do not adequately address technical writing relevant to Civil Engineering. Therefore, a decision was made to focus on courses directly controlled by the Civil Engineering Section that could be used to enhance technical writing skills and IL skills.

Initially, the ABET "a-k" student outcomes were reviewed, and three of them, "g", "i", and "j" were selected for their reference to some components of IL; these

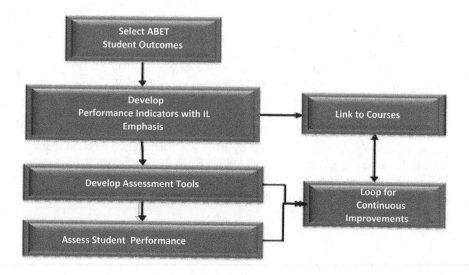

FIGURE 4.1   Process of linking IL to ABET student outcome.

TABLE 4.4
IL across the CE Curriculum

Fall Freshmen Year	Spring Freshmen Year
2111 College Composition	2123 Cultural Perspectives
5102 Chemistry I	8115 Macroeconomic Principles
Fall Sophomore Year	**Spring Sophomore Year**
1206 Mechanics of Materials	1302 Materials for Civil/Construction Engineers
8211 Organizational Behavior and Leadership	
Fall Junior Year	**Spring Junior Year**
1309 Environmental Engineering I	1407 Environmental Engineering II
1304 Soil Mechanics and Foundations Design	1319 Transportation Engineering
Fall Senior Year	**Spring Senior Year**
1401 Construction Project Management (CPM)	1402 Civil Engineering Design (CED)
1404 Geotechnical Engineering Design	2393 Morals and Ethics
	2485 Global Studies

ABET "a-k" were recently updated to the current ABET "1–7" student outcomes. Performance indicators for two outcomes (3-3 and 3-7) related to IL were developed and then mapped or linked to courses within the current civil engineering curriculum. Assessment tools were developed and utilized to measure achievement of each performance indicator. An ongoing effort is being made to progressively infuse some components of IL throughout the Civil Engineering curriculum. The outcomes and related performance indicators developed for each of the new ABET student

TABLE 4.5
Performance Indicators for IL Skill Development

ABET Student Outcome	Performance Indicators	Courses	Assessment Tools
ABET 3-3: an ability to communicate effectively with a range of audiences	**3-3-1:** Use appropriate presentation tools and techniques to orally communicate information, concepts, and technical ideas effectively. **3-3-2:** Prepare written documents in standard engineering format to communicate information, concepts, and technical ideas effectively. **3-3-3:** Research information from a variety of sources, utilize information to make engineering decisions/judgment, and produce a technically sound report **3-3-4:** Respond to questions from diverse audiences with justified and well-formulated answers.	**3-3-1:** Soils, CPM, CED **3-3-2:** Soils, CED **3-3-3:** Enviro I, CPM, CED **3-3-4:** Soils, CED	Laboratory reports, technical paper, Personal journal, capstone project report and presentation
ABET 3-7: an ability to acquire and apply new knowledge as needed, using appropriate learning strategies	**3-7-1:** Take the Fundamentals of Engineering exam. **3-7-2:** Use libraries and other appropriate sources to search for information necessary for engineering projects. **3-7-3:** Identify, investigate, and analyze information related to contemporary topics in civil engineering. **3-7-4:** Use modern engineering tools to foster critical thought in analysis and design.	**3-7-1:** FE (Fundamental of Engineering) Review **3-7-2:** Soils, Enviro II, CPM **3-7-3:** Enviro I, CPM **3-7-4:** Steel, Geotech, Recon	FE pass rate, Laboratory reports, technical papers, Journal, Reflection paper, Oral presentation

outcomes and that were linked to the upper-level Civil Engineering courses are presented in Tables 4.5. Student performance is then assessed in each course on the established performance indicators. Details of the assessment of these outcomes/performance indicators are discussed in Chapter 6.

The different assessment tools used to assess students' performance on the performance indicators are presented in Table 4.5. Details of each assessment tool assignment are as follows:

- *Soil Mechanics Technical Paper* – Students prepare a written technical paper on a selected topic in geotechnical engineering according to established technical writing guideline from a professional organization such as ASCE or American Society for Engineering Education (ASEE). Sample guidelines for the technical paper is shown in Appendix 3. Students are also required to give a 20-minute formal presentation to fellow students and faculty.

The papers and presentations are graded consistently using established rubric. The technical writing and presentation aspects are part of a campus-wide "Graduating Class of 1959" competition. Students with the best written and best presentation are selected separately from each major. Then they compete campus-wide for the best overall paper and presentation. The graduating class of 1959 generously established an endowment for the purpose of encouraging excellence in the writing and speaking skills of the future officers enrolled at the Academy. Each academic year, the Academy holds the "Class of 1959" Contest, and the contest has helped reinforce communication skills within the curriculum by infusing writing and public speaking across all the academic disciplines.

- *Environmental Engineering I Water Quality Report* – Students discuss results of water quality testing in relation to their predictions. Students draw appropriate and reasonable conclusions based on the results of the lab.
- *Environmental Engineering I Student Journal* – Students collect articles on a variety of contemporary environmental issues and write about the articles and issues.
- *Environmental Engineering I Climate Change Paper* – Students use a variety of reliable sources to prepare a research paper based on climate change. The objectives of the paper include: (1) To become familiar with the basic scientific principles behind climate change, (2) to become aware of the challenges that remain in predicting the effects of climate change and in developing solutions to the problem, and (3) to recognize and use established scientific sources to develop an informed position on a contemporary issue.
- *Construction Project Management Technical Journal Review* – Contemporary issues specific to the engineering profession or construction industry are explored. Achievement is assessed through individual student research and written analysis of contemporary industry issues as published in technical journals.
- *Construction Project Management Contemporary Issue Presentation* – Students work in teams of two to research, analyze, and present contemporary issues in engineering or the construction industry. This assignment is a follow-up of the individual effort technical journal review assignment specific to contemporary issues. Achievement is assessed through performance during team presentations.
- *Civil Engineering Design (CED) Project Presentation* – Capstone teams are required to present their team experience and project results to a diverse audience of peers, faculty, professional engineers, and clients. Students work in groups of 4–5 on their capstone projects; each student is expected to be the subject matter expert on a particular topic related to the project and complete a technical paper. There is some "open-endedness" to the assignment because the technical topics are determined based on the scope of each project. Team presentations are approximately 20–30 minutes long, and each team member is required to contribute to the presentation. Achievement is assessed through performance during oral presentations scored using a grading rubric.

Grading rubrics were developed to ensure that the important components of writing and IL (such as scope of research, variety of sources, and use of sources) were consistently assessed and evaluated by different instructors. USCGA faculty opted to develop rubrics that would accomplish the goal of communications assessment and improvement. After reviewing initiatives at other institutions, overly complicated rubrics seemed difficult to use, less sustainable, and could discourage faculty members from embracing assessment and development. For assessment tools where a rubric is used, students received the grading rubric together with the assignment to ensure that the expectations of the instructors were known. Details of the assessment using these rubrics are discussed in Chapter 6.

4.5   CAMPUS RESOURCES-PARTNERSHIP WITH THE LIBRARY

Librarians have the primary responsibilities of facilitating access to information resources and providing instructions on how to use those resources. As such, access to the appropriate library resources plays an important role in helping students develop and enhance their IL skills. One challenge cited by Roberts and Bhatt (2007) in building IL in engineering students is to regularly acquaint them with available library resources and how to intelligently utilize them. Therefore, librarians should be active partners with faculty in the effort to promote IL and help students develop the required IL skills.

The mission of the U.S. Coast Guard Academy Library is to support the Academy's educational and training missions by providing quality library services, resources, and facilities to the students, faculty, and staff. Freshman students are required to attend a library skills session where they are introduced to the library research process and shown how to find books and reference materials and use general, multi-subject databases. The information in this session is built upon and expanded during the freshman year English and History classes when students are provided further instruction on search strategies, more specialized databases, and primary sources to complete a specific assignment. Librarians play a very active role in not only providing access to information but also working with faculty to promote IL. In addition to in-person instruction, the library has created research guides for each engineering major as well as guides that are targeted to specific classes. Class-specific sessions can be requested by faculty to provide research instruction on particular subjects. The library's website also includes several subject and course research guides including an engineering research guide.

4.6   CLOSING THOUGHTS

This chapter emphasizes that communication skills should be progressively and consistently incorporated into undergraduate engineering curricula for students to develop and hone the required skills necessary for engineering practice and life-long learning. Information technology continues to advance, and the need for effective communication skills in engineering practice has become even more important. Students need to develop and improve their communication skills in the context of the current and emerging information infrastructure. This has become even more important with the current trends in online access, texting, and instant communications. As a result, engineering educators must ensure that students learn the important steps of evaluating the credibility and appropriateness of sources while properly synthesizing, using, and citing information from several sources. The development of good research and IL skills is vital for students, who are contending with an increasing amount of choices in the range and quality of information resources available to them. The growth of electronic resources – such as electronic books, journals, databases, and websites – has increased the potential for students to learn independently as they are able to access information outside of the physical campus and in the online learning environment; however, access to so much information increases the importance of teaching students how to appropriately find, evaluate, and use reliable information sources.

The Civil Engineering Program at the United States Coast Guard Academy developed specific performance indicators related to IL and linked them to several ABET student outcomes. The goal is to promote IL and communication skills within the current civil engineering curriculum by developing assignments, assessing student achievement of performance indicators, and making improvements as needed based on the evaluation of the assessment data. The assessment tools were kept as simple as possible and woven into existing assessment practices to keep the burden on faculty members to a minimum. Components of IL and critical communication skills were progressively infused throughout the curriculum to enable students to develop and practice the appropriate skills. It is important that faculty take the lead in promoting communication skills in close collaboration with the English Department, Writing Center, and the library staff so that life-long learning skills are valued and supported in the context of academic disciplines. As the Coast Guard Academy develops a more unified approach to weave essential professional skills, such as communications and IL, throughout the 4-year cadet experience, the approach described in this chapter will serve as a valuable example to adopt by others.

REFERENCES

ABET. (2018). *Criteria for Accrediting Engineering Programs.* ABET Engineering Accreditation Commission, Baltimore, MD.

American Library Association. (1989). *Presidential Committee on Information Literacy Final Report,* Chicago, IL. Retrieved http://www.ala.org/acrl/publications/whitepapers/presidential.

American Society of Civil Engineers (ASCE – BOK3). (2019). *Civil Engineering Body of Knowledge: Preparing the Future Civil Engineer.* 3rd Edition, American Society of Civil Engineers, Reston, VA.

Association of College and Research Libraries. (2000). "Information literacy competency standards for higher education." *College Residential News,* Vol. 61, No. 3, pp. 207–215.

Bruce, C.S. (1997). *Seven Faces of Information Literacy.* Auslib Press, Adelaide.

Burkhardt, J.M., MacDonald, M.C. and Rathemacher, A.J. (2003). *Teaching Information Literacy: 35 Practical, Standards-Based Exercises for College Students.* American Library Association, Chicago, IL.

Campbell, S. (2004). "Defining information literacy in the 21st century." *Paper Presented at the World Library and Information Congress: 70th IFLA General Conference and Council,* August 22–27, Buenos Aires, [www.ifla.org/IV/ifla70/papers/059e-Campbell.pdf] (Acc: 2010-10-18).

Cox, C.N. and Lindsay, E.B. (2008). *Information Literacy Instruction Handbook*. Association of College and Research Libraries, Chicago, IL.

Conrad, S. (2017). "A comparison of practitioner and student writing in civil engineering." *ASEE Journal of Engineering Education*, Wiley, Vol. 106, No. 2, pp. 191–217.

Conrad, S. Pfeiffer, T. and Szymoniak, T. (2012). "Preparing students for writing in civil engineering practice." *Proceedings of the American Society for Engineering Education Conference & Exposition*, San Antonio, TX.

Daniell, B., Figliola, R., Moline, D. and Young, A. (2003). "Learning to write: experiences with technical writing pedagogy within a mechanical engineering curriculum." *Proceedings of the American Society for Engineering Education Annual Conference & Exposition*, Nashville, TN.

Doyle, C.S. (1992). *Outcomes Measures for Information Literacy Within the National Education Goals of 1990*. Final Report to National Forum on Information Literacy. (ERIC Document Reproduction Service No. 351 033).

Ford, J.K. and Riley, L. (2003, October) "Integrating communication and engineering education: a look at curricula, courses, and support systems." *Journal of Engineering Education*, Vol. 92, pp. 325–328.

Grassian, E.S. and Kaplowitz, J.R. (2001). *Information Literacy Instruction: Theory and Practice*. Neal-Schuman, New York.

Harichandran, R., Nocito-Gobel, J., Brisart, E., Erdil, N., Collura, M., Daniels, S., Harding, W. and Adams, D. (2014). "A comprehensive engineering college-wide program for developing technical communication skills in students." *ASEE/IEEE Frontiers in Education Conference*, Madrid, Spain.

Jackson, H., Rumsey, N., Daragan, P. and Zelmanowitz, S. (2011). "Work in progress – assessing information literacy in civil engineering." *Proceedings of the 41st ASEE/IEEE Frontiers in Education Conference*, Rapid City, SD.

Jarson, J. (2010). "Information literacy and higher education: a toolkit for curricular integration." *College and Research Libraries News*, Vol. 71, No. 10, pp. 534–528.

Jernquist, K. and Clippinger, D. (2017). "Writing center supports engineering majors with a common language." *ASEE-Northeast Section Conference*. University of Massachusetts Lowell, Lowell, MA.

Moll, M. (2009). "Information literacy in the new curriculum." *South African Journal of Library & Information Science*, Vol. 75, No. 1, pp. 40–45.

National Forum on Information Literacy http://infolit.org/definitions/.

Niedbala, M.A. and Fogleman, J. (2010). "Taking library 2.0 to the next level: using a course wiki for teaching information literacy to honors students." *Journal of Library Administration*, Vol. 50, pp. 867–882.

Owusu-Ansah, E.K. (2003). "Information literacy and the academic library: a critical look at a concept and the controversies surrounding it." *Journal of Academic Librarianship*, Vol. 29, No. 4, pp. 219–230.

Plumb, C. and Scott, C. (2002, July). "Outcomes assessment of engineering writing at the University of Washington." *Journal of Engineering Education*, Vol. 91, pp. 333–338.

Ratteray, O.M.T. (1985). "Expanding roles for summarized information." *Written Communication: A Quarterly Journal of Research, Theory, and Application*, Vol. 2, No. 4, pp. 457–472. (Beverly Hills, CA: Sage Publications.)

Ratteray, O.M.T. (2000–2002). Oswald Ratteray presented the concept of shared responsibilities for information literacy instruction in at least four forums across the Middle States region: (1) Reshaping the Ivory Tower: The Power of Information Literacy, presented at the Annual Meeting of the SUNY Council of Library Directors, April 5–7, 2000; (2) The Spring Meeting of the Congress of Academic Library Directors of Maryland, May 19, 2000; (3) Are Students Really Learning? Faculty/Librarian Collaboration for Accreditation, an address to The Greater New York Metropolitan Area Chapter of the

Association of College & Research Libraries, December 7, 2001; and (4) The Revised Characteristics: What They Mean for Libraries and Assessment, delivered at the Spring Program of the Delaware Valley Chapter of the Association of College and Research Libraries, May 31, 2002.

Reddy, M. (2007). "Effect of rubrics on enhancement of student learning." *Educate*, Vol. 7, No. 1, pp. 3–17.

Riley, L.A., Furth, P. and Zelmer, J. (2000). "Assessing our engineering alumni: determination of success in the workplace." *ASEE Gulf-Southwest Section Annual Conference*, Las Cruces, NM, ASEE.

Roberts, J. and Bhatt, J. (2007). "Innovative approaches to information literacy instruction for engineering undergraduates at Drexel University." *European Journal of Engineering Education*, Vol. 32, No. 3, pp. 243–251.

Rockman, I.F. (2004). *Integrating Information Literacy into the Higher Education Curriculum: Practical Models for Transformation*. Jossey-Bass, San Francisco, CA.

Saleh, N. (2013). "Integrating information literacy in the engineering curriculum: a program approach." *Proceeding of Canadian Engineering Association Conference*, Montreal, Canada.

Swanson, T. (2011). "A critical information literacy model: library leadership within the curriculum." *Community College Journal of Research and Practice*, Vol. 35, pp. 877–894.

5 Professional Ethics

5.1 INTRODUCTION

Good morals, in one form or another, are first taught by parents, family members, and the community as a crucial part in the child development process. These lessons take the form of parents and family members training children to be honorable, truthful, honest, and respectful and to keep their promises. Morals and ethics are deeply woven into the decision-making processes of everyday life; this intertwines moral and ethics into our standard of living. Typically, the standards vary from one society to another. Regardless of what the standards are, they should be harvested and refined through education into an overarching set of guidelines that govern the actions and behaviors within a particular society. The agreed upon standards on how members of a group will interact (morals) have distinct advantages for that group, as it engenders group cohesion, trust, accountability, and an ability to work together to achieve the common goals within that society. The personal ethical values will be revealed and tested by many situations that arise during normal business activities or personal interactions with others. Certain individuals may decide that there is a personal advantage to acting against (immoral behavior) the agreed upon standards to achieve personal advancement. This personal advancement will invariably be at the expense of other group members, and if everyone decided to act immorally, then the group (or society) will collapse. Individual recognition of the advantages of morality within the group and thereby deciding to abide by those standards is the first step in becoming an ethical individual. Ethics can be viewed as a system of moral principles defined by a society and utilized in guiding the decision-making process in serving the needs of society. Therefore, the goal of setting ethical standards allows individuals to improve their decision-making process on morally complex issues in ways that respect and benefit society. Within the engineering community, "Engineering Ethics" is the study of moral issues and decisions confronting individuals and organizations engaged in engineering work. It is the engineer's responsibility to gain public trust by designing and building infrastructure that emphasizes the safety, health, and well-being of the public, and beneficial to society. This responsibility also requires understanding the importance of professional ethics and its application in any engineering endeavor.

In addressing engineering ethics in undergraduate education, a clear distinction must be made between moral behavior and professional ethics within different cultures and professional settings. To accomplish this, the curricula of engineering programs should be infused with a more global perspective on ethics and its relevance in the practice of engineering throughout the world. Professional ethics education will not by itself reduce unethical action or practice. The awareness of proper conduct and the empowerment of individuals to challenge practices are critical outcomes of a professional ethics educational program that will contribute significantly to the

DOI: 10.1201/9781003280057-5

ongoing preservation of ethical behavior in any profession. A comprehensive professional ethics education will support individuals to become critically aware and scrutinize practices around them rather than becoming enculturated with existing norms, practices, and values of ethical professional behavior.

The focus of this chapter is on how students develop as ethical leaders with an appreciation for cross-cultural issues by infusion of professional ethics topics in the undergraduate civil engineering curriculum, in core courses, and in co-curricular activities. Examples of how student development can be achieved both inside and outside the classroom to prepare engineering leaders who can ethically meet professional challenges are discussed.

5.2   PROFESSIONAL SOCIETIES EFFORTS IN ETHICS EDUCATION

Professional societies develop policies and standards including codes of ethics to help govern members' behavior, responsibilities, and accountability. The codes are used to define accepted practices of their members toward each other and toward the public. Professional ethics encompass how people work in a professional setting by following expected standards and goes beyond simple right and wrong decisions. Fundamental to all codes is fairness, respect, trustworthiness, integrity, as well as public health and safety. The National Society of Professional Engineers adopted a general code of ethics for engineers in all disciplines that was updated in 2019 (NSPE, 2019). This code consists of six fundamental canons that engineers should fulfill in their duties. In summary, these canons state that an engineer should always put public safety first, only perform services in their area of expertise, be faithful and avoid deception while conducting themselves in an honorable, responsible, and ethical way to enhance the reputation of their profession. It is difficult to change the morals and ethics of a person; however, it is possible to expose or teach a person to think ethically as a professional through education. Engineering educators have used tools to emphasize the importance of ethics that include exposing students to professional engineering codes of ethics and engineering standards. The challenge is to ensure that students are not only aware of the codes of ethics and standards of behavior but know how to interpret and apply them in real-world situations. Educators can help students understand and internalize the core values of safety and environmental protection by situational awareness of taking action to protect society. This could be achieved by encouraging students to develop a keen awareness of the consequences of their decisions (both immediate and long-term) as well as the associate risks to safety and well-being of the public. For example, Johnson (2017) presented four goals that must be achieved in teaching engineering ethics by focusing on what can and should be taught and how best to teach it:

- Make students aware of what will be expected of them in their work as engineers,
- Sensitize students to ethical issues that might arise in their professional practice,
- Improve students' overall ethical decision-making and judgment, and
- Motivate and inspire students to behave ethically.

The American Society of Civil Engineers (ASCE), in response to feedback from practicing engineers and other civil engineering professionals, articulated new standards for civil engineering programs to meet in regard to professional skills that are included in the professional component of the Body of Knowledge (BOK3, 2019). This professional component includes proficiency in communications, public policy, business and public administration, globalization, leadership, teamwork, personal conduct, lifelong learning, and professional and ethical responsibilities. The BOK3 suggests that undergraduate students begin experiencing the civil engineering role in society and the environment which should "naturally lead to the importance of professional competency and the need for ethical behavior." It has been suggested to introduce students to professional standards and codes of ethics early in their undergraduate education. Over time, this should evolve into applying ethical codes and standards in professional conduct and practice to determine the appropriate action after analyzing the specific circumstances and possibly weighing conflicting interests. The BOK3 also suggests that learning should be accomplished across the curriculum and in selected co-curricular and extracurricular activities. Furthermore, the engineering accreditation organization, ABET recognized the needs for these professional skills and requires all civil engineering programs to demonstrate that their graduates have an "ability to recognize ethical and professional responsibilities in engineering situations and make informed judgments, which must consider the impact of engineering solutions in global, economic, environmental, and societal contexts" (ABET, 2018). This ABET requirement does not specify particular formulation of goals within these broad categories nor particular strategies to meet the requirements, except to say that there should be a general education component that complements the technical content of the curriculum. ABET does not provide any structured methods of implementing this requirement. It is up to the engineering educators to develop a robust plan to teach students the ethical understanding and decision-making skills society requires of the engineering industry.

Professional codes of practice should be dynamic and regularly revised to address any changing behavior and culture within the professional industries. In this regard, ASCE enacted a new set of professional ethics in 2020 that are streamlined to better safeguard society and the civil engineering profession. The foundation of the new ASCE Code of Ethics (COE) consists of five stakeholders (society, environment, profession, client/employee, and peers) in order of priority as shown in Figure 5.1. More details about this can be found on the ASCE's website (www.asce.org). A copy of the new ASCE COE is included in Appendix 1.

5.3   ACADEMIC EXAMPLES OF ADDRESSING ENGINEERING ETHICS

Ethics across the curriculum is important to demonstrate that ethics is not simply an academic exercise, limited to college campuses, but that ethics is a part of the professional life of an engineer.

Although using codes of ethics may be an informative and effective framework, memorizing what the codes are and promising to follow them is not adequate preparation for engineering practice. Codes should be used as a basis for instruction only if

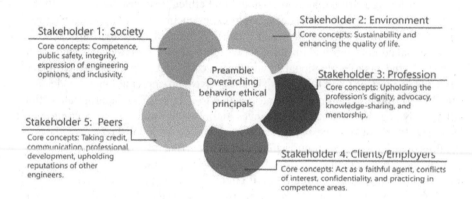

FIGURE 5.1   Foundation of the 2020 ASCE Code of Ethics. (With permission from ASCE.)

educators focus on and show the value of these codes by emphasizing to students how they relate to their future career. Davis (1991) as well as Davis and Feinerman (2012) suggested that there should be a minimum of four desired outcomes when teaching ethics courses: (1) an increased ethical sensitivity, (2) an increased knowledge of relevant standards of conduct, (3) an improved ethical judgment, and (4) an improved ability to act ethically. An effective strategy is for engineering educators to engage students and attempt to interact with them through role-playing and discussions. Modifying a purely technical assignment by including a component that requires student to address ethical aspects can be another effective tool. Carpenter et al. (2014) discussed the need to prepare engineers to reason through and act appropriately on the ethical dilemmas they will experience as professionals. The investigation evaluated different institutional approaches for ethics education with a goal of better preparing students to be ethical professionals. The project included visiting 19 institutions and collecting data from nearly 150 faculty and administrators and more than 4,000 engineering undergraduates completed the survey. The project findings suggest that:

- co-curricular experiences have an important influence on ethical development,
- the quality of instruction on engineering ethics is more important than quantity of curricular experience,
- students are less likely to be satisfied with professional ethics instruction when they have higher ethical reasoning skills,
- the institutional culture affects how students behave and how they articulate concepts of ethics.

Carpenter et al. (2014) also suggest that an important factor in developing students' ethical reasoning ability is exposing them to curricular experience that require deep cognitive processing about ethical issues at the appropriate level of Bloom's Taxonomy. They concluded that engineering should focus on improvements in instructional methods and instructional environments, as well as ensuring that individuals responsible for delivering ethical content are capable and motivated instructors. They also recommended promoting extracurricular and co-curricular involvement through service programs and professional organizations. Institutions should establish and communicate clear behavioral expectations across administration, faculty, staff, and students such that the culture of the institutions promotes ethical development. Carpenter et al. concluded that the curricular foundation is in place, but that institutions need to improve their curricular and co-curricular offerings to facilitate ethical development of students and fulfill ASCE Body of Knowledge outcomes.

Colby and Sullivan (2008) offered strength and weaknesses analysis after observational site visits to 11 undergraduate engineering programs in seven engineering schools across the United States of America to determine how undergraduate engineering programs support students' ethical development. They concluded that a good framework to facilitate student learning of professional responsibilities is to include professional code of ethics into the curriculum. Codes of engineering ethics make it clear that complying with the codes requires a great deal of knowledge, ethical sophistication, interpersonal skill, personal integrity, and sound judgment. The codes also emphasize the important fact that engineering competencies, a commitment to high-quality work, and a drive to keep learning are inseparable from other aspects of professionalism. Furthermore, they suggested teaching for professional responsibilities benefits from pedagogies of guided practice, learning by doing, with models of expertise and informative feedback on performance. Engineering faculty were encouraged to integrate ethics and professional responsibilities more thoroughly into the curriculum. Faculty might contribute by highlighting ethical issues in teamwork or design, drawing attention to the potential of engineering work to enhance human welfare or the environment, drawing out the meaning and long-term impact of technology, linking intellectual and academic integrity in the workplace, and the like.

Another approach to teaching ethics that has been successfully implemented in several institution is the "Giving Voice to Values" (GVV) approach developed by Prof. Mary Gentile. This approach is based on encouraging students to develop an "ethical memory muscle" so they are prepared when faced with ethical dilemmas. It involves strategizing, preparing, and practicing so students can effectively act on their values. The main objective of GVV is not to persuade students to be more ethical but to raise the odds of them effectively and successfully acting on their values through rigorous practice. This approach to values-driven leadership is built around preparing for and practicing values-based action, and it answers the following questions (University of Virginia – Darden school of Business):

- How do I learn to act on my values?
- What would I say and do?
- How can I be the most effective in acting on my values?

It has been challenging for most academic faculty to teach ethics because of a lack of professional background to adequately address the complexities of moral and ethical behavior. The Center of Engineering Ethics and the Online Ethics Center have developed resources to assist the engineering industry in the effort of promoting an appropriate level of instruction and development in ethics instructions. "The online Ethics center provides students, educators, and practicing engineers and scientists' resources for understanding and addressing ethical problems." The Ethics CORE (Collaborative Online Resource Environment) Resources project is an Internet portal supporting ethics education in science, social science, engineering, and math. It was developed by the National Center for Professional and Research Ethics at the University of Illinois-Urbana Champaign. The online environment consists of tools such as searching, developing, and contributing resources, collaborative workspaces, discussion areas, wikis and blogs as well as essays on teaching and pedagogy, videos, online courses, and links to other online resources. The portal can be accessed at http://nationalethicscenter.org/. For example, "The Practice of Ethics in Classroom Teaching" is a video showing ways to support student learning through ethical faculty behavior. All members of the engineering education community are encouraged to participate, whether by contributing resources or feedback, by actively participating in collaborative groups, or by using resources to enhance their teaching. Other online resources for engineering ethics education are also available, primarily in the form of case studies that can be used in classes. Some examples include:

- National Institute for Engineering Ethics, Cases from the National Society of Professional Engineers Board of Ethical Review: http://www.niee.org/cases/
- Online Ethics Center, http://onlineethics.org, is a product of the National Academy of Engineering (NAE). It includes resources for responsible research, case studies, professional codes and guidelines, annotated bibliographies, and a community of practitioners.
- Texas A&M Engineering Ethics: http://ethics.tamu.edu/
- The Ethics Education Library at the Center for the Study of Ethics in the Professions at IIT: http://ethics.iit.edu/node/62 (with other information at their site http://ethics.iit.edu).

The National Academy of Engineering (NAE, 2016) compiled a report on how 25 exemplary academic institution infused ethics into the development of engineers in their curricula. In preparing the report, the Center for Engineering Ethics and Society reviewed several engineering ethics education activities from different institutions with the goal of selecting and widely disseminating those that may serve as exemplars for broader adoption and adaptation. This report aims to raise awareness to the variety of exceptional programs and strategies for improving engineers' understanding of ethical and social issues and provides a resource for those who seek to improve the ethical development of engineers at their own institutions (NAE, 2016). The report also classified three challenges and provides suggestion on addressing them:

- *Student Challenges* – a lack of interest among students,
- *Faculty Resistance* – resistance and reluctance from faculty to teach ethics, and
- *Topical and Pedagogical Challenges* – a lack of consensus on important topics and methods for incorporating ethics in an already overstuffed curriculum.

Examples on how ethic is addressed by some of the academic programs included in the NAE Exemplar are summarized in Table 5.1.

TABLE 5.1

Examples of Academic Approaches to Infusing Ethics into Engineering Programs

Institution	Description	Outcomes
Georgia Institute of Technology	Problem-based professional ethics course for juniors and seniors. Course format includes the use of open-ended practical scenarios in which students consider the ethical implications for each option in terms of basic moral values.	• Students develop the ability to identify, respond to, & reflect on ethical values. • Students develop an understanding of ethics in engineering design thinking and engineering projects. • Students develop the ability to critically think about a range of ethical situations. • Students develop an appreciation of ethical implications and develop alternative approaches.
Kansas State University	The course focuses on the ethical responsibilities of being an engineer. Students are required to interview practicing engineers about ethical challenge(s) and develop an engineering ethics workshop for their peers.	• Students develop the ability to analyze an ethical engineering dilemma. • Students develop an appreciation for different ethical concepts and approaches. • Students develop a personal code of ethics as a roadmap for their professional career. • Students develop the courage and resilience to champion future ethical solutions.
Northeastern University	This course uses biomedical engineering ethics case studies to address potential ethical issues for the entire life cycle of biomedical products. The case studies are collectively designed for students by practicing engineers, policy experts, business professionals, and ethicists.	• Students develop appreciation for ethical issues from multiple angles. • Students develop an understanding of the indirect impacts of professional activities on society. • Students enhance their critical thinking skills in addressing the life-cycle issues and environmental ethics of product development.

(Continued)

TABLE 5.1 (*Continued*)
Examples of Academic Approaches to Infusing Ethics into Engineering Programs

Institution	Description	Outcomes
University of Pittsburg	Course in Biomedical Engineering that uses student-authored case studies to foster discussion and reasoning in ethics. Students develop their own case study by recognizing "ill-defined" every day practical scenario. Students serve as peer-reviewers for in-class group projects.	• Students learn how to identify, frame, and resolve multidisciplinary ethical issues in professional practice. • Promotes teamwork and a good learning environment for group discussions.
University of Virginia	Senior Thesis activity that challenges students to integrate social and ethical analysis with engineering by building on an understanding of the relationship between engineering, technology, and society.	• Students develop competence in ethical awareness and analysis. • Fosters development of communication skills. • Students develop an understanding of the relationships between science, technology, and society and the implications of these relationships for engineering practice.
Virginia Polytechnic Institute & State University	This course addresses the ethical responsibilities of engineers to engage with the public and other stakeholders. The approach is to use a tool (learn to listen) for morally engaged engineering practice.	• Students develop ethnographic listening to diverse public concerns to understand the impact of proposed alternatives to complex engineering projects. • Fosters the development of ethical decision-making. • Students learn how to engage local experiences, knowledge, values, understand professional power and technical expertise.
Worcester Polytechnic Institute	First-year general education course that teaches engineering content in a complex social environment where ethical questions are part of engineering practice. Course is taught by a diverse faculty from humanities, social science, and engineering. Format includes role-playing activities.	• Students learn and practice engineering content through interactive role-play as engineers, business, scientists, or laborers. • Students gain an appreciation to research and debate macro ethical questions. • Students develop communication skills and function on multidisciplinary teams. • Students learn to address complex social problems with creativity, cross-cultural communication skills, and an appreciation for diverse viewpoints.

Source: Adapted from NAE (2016).

5.4 ETHICS ACROSS THE CURRICULUM AT USCGA

The United States Coast Guard Academy's (USCGA) mission is: "To educate, train and develop leaders of character who are ethically, intellectually, and professionally prepared to serve their country and humanity." Coast Guard personnel constantly face challenges in mission operations involving diverse cultures, and to meet these challenges, they must regularly act ethically and exercise competent leadership. USCGA articulates and instills the Coast Guard core values that include *Honor, Respect, and Devotion to Duty*. These core values are the framework for the professional conduct and ethical responsibilities that the Coast Guard expects. USCGA fosters ethical leader development and global awareness through a breadth of required core courses in the humanities, science, engineering, mathematics, professional maritime studies, organizational behavior, management, leadership, and law. Student achievement of specific performance indicators related to ethics and global issues is assessed regularly, and curricular improvements are implemented as needed. Co-curricular programs, such as the "Annual Ethics Forum" and the "Professional Conduct & Risk Management Workshops," support the Academy's mission in preparing cadets to be leaders in a dynamic and ever-changing world. The combination of core courses, major-specific engineering courses, and co-curricular activities provides students the opportunities to develop leadership and professional ethical conduct required for the engineering practice.

Ethics instruction is embedded into all the required academic courses at USCGA. It is especially emphasized in the junior and senior level courses. As juniors, civil engineering students take Morals and Ethics (3 credits) and Criminal Justice (3 credits) courses. As seniors, they continue their study of law with Maritime Law Enforcement (3 credits). The Morals and Ethics course includes two main components: (1) *Ethical theories*, both historical and contemporary, along with the arguments for and against them; and (2) *Applied Ethics*, both in general and utilizing case studies within a specific field. Throughout the semester, students examine a range of philosophical views regarding what makes actions right or wrong, characters good or bad, help students reinforce their decision-making abilities, develop their own moral voice, and appreciation for the place of reasoned argument in the treatment of ethical problems. The objectives of the Morals and Ethics course are shown in Table 5.2. These course objectives are aligned with the Academy's shared learning outcomes, shown in column 1 of Table 5.2. The shared learning outcomes (discussed in Chapter 4) are observable results that define the unique qualities, character, and knowledge of USCGA graduates (2010). Students also study and explore the basic legal concepts in Criminal Justice and Maritime Law Enforcement and learn specifically about the United States civilian and military criminal justice system and the legal issues associated with the Coast Guard's law enforcement mission in the maritime environment.

Within the Civil Engineering curriculum, ethical and global issues are woven into the courses progressively. To ensure continuous assessment and improvement of coverage in these areas, USCGA's well-established ABET assessment system is used to evaluate student progress. This progress is assessed by monitoring student performance on performance indicators associated with certain ABET student outcomes.

TABLE 5.2

Morals and Ethic Course Objectives and Shared Learning Outcomes

USCGA Shared Learning Outcomes	Course Objective
Communication	Read and understand a variety of writings in moral philosophy.
	Participate in discussions of these writings by listening critically to oral arguments and asking thoughtful questions.
	Write clear, concise, persuasive, and grammatically correct papers on a variety of ethical issues and theories.
Acquire, integrate, and expand knowledge	Understand a variety of ethical theories and the main arguments associated with major issues within a field of applied ethics; integrate these theories and arguments into a moral framework that is related to the Coast Guard's Core Values and lends itself to continued expansion.
	Access information regarding an ethical and moral issue; locate and evaluate articles on ethical and moral subjects reliably.
Critical thinking	Recognize conflicts in and between various ethical theories and moral views and use reasoned arguments to support the resolution of these conflicts.
	Use reasoned argument, critical analysis, and problem-solving skills to evaluate moral views, clarify moral problems, and minimize moral disagreements.
Leadership	Develop your own moral views and relate them to the Coast Guard's Core Values through honest, realistic, and constructive self-evaluation; articulate and support your moral views both orally and in writing.
	Understand the complexity of moral life, appreciate the diversity of moral views which results from this complexity, comprehend that there may be no simple answer to an ethical problem, and respect the diversity of values held by reasonable people.

For example, the performance indicators for two former ABET Student Outcomes (3f and 3h) that were used up to 2018 to assess ethics in the civil engineering curriculum are as follows:

ABET 3f: "an understanding of professional and ethical responsibility" *is evaluated with three specific performance indicators.*

- 3f-1 – Articulate importance of professional code of ethics.
- 3f-2 – Identify ethical dilemmas and propose ethical solutions in accordance with professional code of ethics.
- 3f-3 – Investigate a given engineering project and articulate ethical issues.

ABET 3h: "the broad education necessary to understand the impact of engineering solutions in a global, economic, environmental, and societal contexts" *is addressed with two performance indicators.*

- 3h-1 – "explain the economic, social and global aspects of engineering solutions" and
- 3h-2 – "discuss the environmental implication of engineering solutions"

Faculty members crafted assignments and developed rubrics related to these performance indicators to ensure student development in ethical and global issues relating to Civil Engineering. These two former ABET Student Outcomes (3f and 3h) were combined into the new ABET (2018) Student Outcome 3–4: "an ability to recognize ethical and professional responsibilities in engineering situations and make informed judgments, which must consider the impact of engineering solutions in global, economic, environmental, and societal contexts." The performance indicators used to assess Student Outcomes "3f and 3h" were combined into four performance indicators to assess the new ABET 3-4 Student Outcome, namely,

- 3-4-1 – Articulate importance of professional code of ethics.
- 3-4-2 – Investigate engineering problems or case studies, articulate ethical issues, and propose solutions in accordance with professional code of ethics.
- 3-4-3 – Explain the economic, social, and global aspects of engineering solutions.
- 3-4-4 – Discuss the environmental implication of engineering solutions.

Coverage of ethics within the Civil Engineering Program is currently addressed in detail as part of the Civil Engineering Design course. Students in this capstone design course apply a variety of knowledge from a broad range of technical, managerial, and humanities coursework to produce solutions that consider the economic, socio-political, ethical, and environmental aspects of real-world problems. Students produce engineering calculations, construction drawings, project schedules, cost estimates, and any other necessary project-specific documents. In addition, students communicate the results of their capstone project via a final report and presentation to their client. A component of the course is to write leadership essays and research and present an ethical scenario from ASCE's "A Question of Ethics" case study archive. The case studies are related to the ASCE Code of Ethics. Each group is allotted 15 minutes to present their Ethics Case Study and the team facilitates a short in-class discussion.

The objective of the ethics case studies is to present relevant engineering ethical situations in the classroom to stimulate discussion of ASCE Code of Ethics and critical thinking. Students were required to research, identify, review case studies in relation to ASCE Code of Ethics, and present their findings to the entire class. Examples of case study presentations in 2018–2019 academic year based on the previous ASCE Canons are shown in Table 5.3.

Professional ethics is consistently emphasized throughout the civil engineering curriculum. Some examples of additional opportunities in other course that address professional ethics are highlighted below:

- Case studies, practical examples, or demonstrations are used where appropriate in various civil engineering courses. For example, in the "Structural Analysis" course, the instructor developed a "professional practice moment" in which students take turns presenting a current event (including ethical conduct) related to structural engineering during the last five minutes of each class.

TABLE 5.3

Examples of Ethics Case Study Presentations (2018–2019)

Presentation Topic	ASCE Canon Addressed
Ensuring the safety, health, and welfare of the public	ASCE Canon 1: "Engineers shall hold paramount the safety, health and welfare of the public and shall strive to comply with the principles of sustainable development in the performance of their professional duties."
An engineer's misrepresentation of his credentials and dishonesty	ASCE Canon 2: "Engineers shall perform services only in areas of their competence."
Engineers who gave false geotechnical information	ASCE Canon 3: "Engineers shall issue public statements only in an objective and truthful manner."
The proper use of professional credentials	ASCE Canon 4: "Engineers shall act in professional matters for each employer or client as faithful agents or trustees, and shall avoid conflicts of interest," and
	ASCE Canon 5: "Engineers shall build their professional reputation on the merit of their services and shall not compete unfairly with others."
Fraud	ASCE Canon 6: "Engineers shall act in such a manner as to uphold and enhance the honor, integrity, and dignity of the engineering profession and shall act with zero tolerance for bribery, fraud, and corruption."
Employer's responsibility to employees	ASCE Canon 7: "Engineers shall continue their professional development throughout their careers and shall provide opportunities for the professional development of those engineers under their supervision."

- In the "Geotechnical Engineering Design" course, students are engaged in the process of forensic investigation and evaluation by reviewing four case studies. One of the case studies involves an ethical dilemma. The use of case studies provides students with opportunities to make the connection between theory and real-life application of engineering principles and concepts.
- In the "Environmental Engineering I" course, a case study is presented involving exceedance of pollutant limits. Students evaluate the situation from multiple perspectives and relate the issues to the Engineering Code of Ethics. Their progress is evaluated using a rubric linked to performance indicators. Students also research and develop presentations to the entire class on various Superfund sites around the country. This is effective in showing a variety of problems and remediation technologies as well as the legal, ethical, and societal issues involved in identifying and cleaning up hazardous waste sites.

5.4.1 ETHICS LUNCHEONS AND FORUM

Engineers must be familiar and committed to the ethical code of the profession. To emphasize the ethical professional engineering conduct, a combined ethics manual was developed for all the engineering disciplines (Civil, Electrical, Mechanical,

Engineering Ethics and the Law: When the Law and the Code conflict.

Objectives:

1. Understand the difference between an ethical and legal requirement.
2. Discuss an engineer's ethical duty when the law and a code of ethics conflict.
3. Participate in a facilitated discussion on how you would convince a boss that the law is not enough.
4. Compose an essay that incorporates the group discussion that effectively answers a related ethical question.

Notes:

Lifeboat Requirements (from the reading):
- British Board of Trade required 16 lifeboats
- TITANIC had required 16 lifeboats plus an additional 4 collapsible
- Original design called for 32 boats

Engineering innovations developed for the TITANIC (from the reading):
- Hull was divided into watertight compartments
- Watertight doors were controlled from the bridge
- Designed to withstand flooding in two compartments

Other Notes:
- Contract was "Cost-Plus" basis – cost of extra life-boats would not have impacted the engineering/shipbuilding company's profits. However, the article says nothing about if TITANIC's owners could afford the extra costs?
- One of the TITANIC's designers actually wanted to make the hull thicker but was
- overridden by the ship owners/engineering firm because it would have made the ship too
- heavy & therefore too expensive to run.1

Talking Points:

1. Given that the R.M.S. TITANIC actually had more than the legally required number of life-boats, did the TITANIC's engineers have an ethical duty to ensure there were enough life-boats on board for the number of passengers and crew? Did the TITANIC's owners? Why or why not?
2. Would your answer change if you found out (hypothetically) that the TITANIC's owners pressured the British Board of Trade's decision not to increase the number of lifeboats?
3. How do you think that the belief that the TITANIC was unsinkable (i.e., the engineers had designed the ship as safe as the technology allowed) impact this ethical responsibility?
4. How would you convince your boss of the need to exceed the legal requirements for a design – especially if there were substantial costs associated with the redesign?
5. This is a case where the code of ethics forces the engineer to "overdesign" or "improve upon" what is legal – so the "ideal" solution is both legal and ethical. Can you think of cases where:
   - what is ethical would be illegal?
   - what is legal would be unethical?

FIGURE 5.2   Example of an ethics lunch talking points.

and Naval Architecture) offered at the USCGA. The manual also included examples of real-life documentation of ethics cases. These cases were discussed during a series of "Ethics Luncheons." This approach was used for several years as part of the civil engineering curriculum. Students were expected to attend several luncheons during which cases involving professional ethics are discussed in groups with a faculty member as moderator. These luncheons were incorporated as a part of the senior year curriculum where each student must attend at least four ethics luncheons and submit a deliverable for each event to fulfill the course requirements. Faculty mentors were given talking points guidelines for each discussion, an example of which is shown in Figure 5.2. The focus of these ethics discussions was to have the students think about the choices they would make in each situation or case study. This changes the ethical lessons from a purely theoretical discussion to one requiring the students to develop, apply, and defend their moral values and choices. The faculty felt this approach of engaging a small group of students in an ethical case study was an appropriate approach to teach engineering ethics. Student written comments such as "It made me realize how sometimes I will be alone in one of those arguments and will have to put my ethical values to the test" validated faculty assumptions. Students learned that it is not always the choice made but the reasoning behind the choice that is most difficult to develop. The use of the ethics luncheons had mixed reviews by both students and faculty. Most of the faculty felt that a significant number of students failed to adequately prepare before the small group discussions. The civil engineering faculty thought that most of the cases studies were too general, and they did not specifically focus on the professional ethics civil engineers generally face in their career. Currently, this practice of combining all engineering students during luncheons has been replaced with more formal ethics instructions in the upper-level courses as previously discussed.

Beginning in the spring of 2021, a new one credit ethics seminar was included in the engineering curriculum. This course is taken by all engineering students and builds upon ethical theory taught in the introductory ethics course, *Introduction to Moral & Ethical Philosophy*, and applies this theory specifically to engineering through case studies and guest speaker discussions. Historical cases are taken primarily from scholarly literature on engineering ethics with the primary goal of familiarizing students with Engineering Codes of Ethics. Students gain knowledge and insight into the issues of ethics in engineering. The course provides preparation for real-world practice of engineering by exposing students to ethical dilemmas that engineers could face through case studies and discussions of resolutions of the ethical dilemmas posed. Assessment tools include student journals, journal peer reviews, and case study presentations.

The Academy also sponsors a 1-day "Annual Ethics Forum" and requires all students to attend at least three or four sessions. The Forum starts with a keynote speaker the night before followed by a daylong conference devoted to ethics. The small group sessions typically cover a diverse array of topics including law, science, engineering, political, military, and financial ethics. Expert speakers from all over the United States are invited to present various sessions and share current ethical cases and their experience with students.

5.4.2 Extra Curriculum Activities

Participation in non-academic activities outside of the classroom is part of the USCGA culture; students make the time to fit these extracurricular activities into their busy schedule. Students are also encouraged to participate in recognized social clubs. Each student is required to complete at least 8 hours of community service per semester. Students rise to the challenge of developing and practicing their professional skills by training and mentoring other cadets, leading a student organization, actively participating in sports, being active volunteers in the community, and organizing and running activities for the Corps of Cadets and the Academy. All of these activities provide opportunities for student development of professional skills including professional conduct and ethics. For example, various groups of cadets participate in planning the annual Ethics Forum, Parent's Weekend, Homecoming, Research Symposium, and Graduation. Throughout these activities, students practice teamwork, goal setting, diversity appreciation, tolerance, conflict management, and communication. Students are expected to manage their time effectively to meet the academic and non-academic rigors. They begin to see the "big picture," develop a sense of purpose, serve others, and be motivated to succeed. Students develop effective study habits, time management, leadership, and professional skills that serve them well during their career in the Coast Guard and in the civil engineering profession.

USCGA civil engineering students are also actively involved in professional organizations such as ASCE, Society of American Military Engineers, Society of Women Engineers, and the Moles (an organization of individuals engaged in heavy construction). Students are encouraged to take on leadership and mentorship roles in student chapters of these professional society. Through these organizations, students interact with practicing professionals who are willing to share their experience with them. Such experiences typically involve mentorship by experienced engineers who will pave the way to successful professional practice.

5.5 CLOSING THOUGHTS

The focus of this chapter was on how students develop skills and appreciation to become ethical leaders by infusing professional ethics topics in the undergraduate civil engineering curriculum, in core courses, and in co-curricular activities. Examples of how student development can be achieved both inside and outside the classroom to prepare engineering leaders who can ethically meet professional challenges were also discussed. At the Coast Guard Academy, developing Coast Guard Officers to serve the nation is an all-hands effort during the 4 years or 200 weeks journey. The academic, military, and athletic realms of student life are tied together through a common purpose to develop ethical leaders and professionals through a host of co-curricular activities, extracurricular experiences, and direct engagement with ethical, professional, and global issues in and out of the classroom. In particular, the Civil Engineering faculty members developed assignments and other curriculum experiences to advance cadet development in professional ethics and cross-cultural issues. The faculty actions are guided by the ABET assessment framework, ASCE Code of Ethics, and the Academy Honor's concept. These activities make students aware of what will be expected of them in their professional work, improve students overall ethical decisions, and inspire students to always behave ethically. The success of the USCGA approach to ethics, professionalism, and global awareness rests on the infusion of these competencies in every aspect of student life throughout their 4-year experience.

REFERENCES

ABET. (2018). *Criteria for Accrediting Engineering Programs.* ABET Engineering Accreditation Commission, Baltimore, MD.

American Society of Civil Engineers. "American society of civil engineers code of ethics." http://www.asce.org/code_of_ethics/.

American Society of Civil Engineers (ASCE – BOK3). (2019). *Civil Engineering Body of Knowledge, Preparing the Future Civil Engineer,* Third Edition, ASCE Press, Reston, VA.

Carpenter, D., Harding, T., Sutkus, J. and Finelli, C. (2014). "Assessing the ethical development of civil engineering undergraduates in support of the ASCE body of knowledge." *ASCE Journal of Professional Issues in Engineering Education & Practice,* Vol. 140, p. A4014001.

Colby, A. and Sullivan, M. (2008). "Ethics teaching in undergraduate engineering education." *Journal of Engineering Education,* Vol. 97, No. 3, pp. 327–338.

Davis, M. (1991). "Thinking like an engineer: the place of a code of ethics in the practice of a profession." *Philosophy and Public Affairs,* Vol. 20, pp. 150–167.

Davis, M. and Feinerman, A. (2012). "Assessing graduate student progress in engineering ethics." *Science and Engineering Ethics*, Vol. 18, No. 2, pp. 351–367.

Engineering, N. A. (n.d.). "Online ethics center." Retrieved from www.onlineethics.org.

Ethics CORE Resources. "The practice of ethics in classroom teaching." https://source.nationalethicscenter.org.

Godfrey, D., Taylor, T., Fleischmann, C. and Pickles, D. (2008) "Teaching engineering ethics in a multi-disciplinary environment." *ASEE Annual Conference*, Pittsburgh, PA.

Johnson, D. G. (2017). "Can engineering ethics be taught?" *The Bridge, National Academy of Engineering*, Vol. 47, No. 1, Spring 2017, pp. 59–64.

National Academy of Engineers. "Ethics codes and guidelines." Online Ethics Center for Engineering. http://www.onlineethics.org/Resources/ethcodes.aspx.

National Academy of Engineering, Infusing Ethics Selection Committee. (2016). *Infusing Ethics into the Development of Engineers, An Exemplary Education Activities and Programs.* National Academies Press, Washington, DC.

National Society of Professional Engineers. (2019). *NSPE Code of Ethics for Engineers.* Alexandria, VA. www.NSPE.org, Revised July 2019

Pridmore, A., and Hoke, T. "Engineering the future: 2020 code of ethics." *ASCE Presentation.* ASCE online webinar (AWI 110520), asce.org, Reston, VA.

United States Coast Guard Academy (2010). *United States Coast Guard Academy Organization and Regulations Manual.* Academy (USCGA), New London, CT.

University of Virginia, Darden School of Business. Website IBIS Initiatives - Giving Voice to Values (GVV) | UVA Darden School of Business (virginia.edu), accessed 28 July 2021.

6 Assessment of Student Learning

6.1 INTRODUCTION

Assessment is the process of evaluating evidence of student learning with respect to specific learning goals. Developing effective assessment methods that focus on student learning and intellectual development is a challenging task because of the diversity of the changing student cohorts, abilities, and learning styles. Therefore, aligning course activities with learning outcomes is critical in implementing an effective assessment strategy. Assessment should be based on what students are learning; it should be systematic with a focus on improving the quality of student learning through appropriate programs, curricula, and pedagogy. The assessment process must also be well documented by promoting the collection of evidence on student learning and interpretation of the information. This system of assessment should provide feedback to students about their learning, and to teachers about their instruction and evidence to support teachers' judgments about grading. There are varying definitions of assessment in the literature; however, according to Spurlin et al. (2008), an effective assessment process should consist of the following four steps:

1. Development of clearly articulated written statement of expected learning outcomes.
2. Design of learning experience that provides intentional purposeful opportunities for students to achieve these learning outcomes.
3. Implementation of appropriate measures of student achievement of key learning outcomes.
4. Use of assessment results to improve teaching and learning.

For most engineering programs, the assessment process is greatly influenced by the need to meet accreditation requirements. Therefore, the strategies tend to focus on the performance of students in fulfillment or achievement of "student outcomes." In the United States, accreditation requirements are set by ABET, Inc (formerly known as Accreditation Board for Engineering and Technology). The focus of the assessment process within engineering undergraduate programs then becomes determining the type and amount of data on student performance that should be collected to satisfy ABET requirements, ensure continued accreditation, and foster continuous program improvement. Academic institutions continue to endeavor to make the task of data collection and the overall assessment process more manageable. This process becomes especially challenging with the dynamic nature of some of the ABET criteria. The importance of accreditation is valued by colleges and universities across

DOI: 10.1201/9781003280057-6

the United States and internationally because it assures the quality of engineering and technology programs as well as facilitates the improvement of those programs. Furthermore, a degree from an ABET accredited institution is one of the requirements to start the process for licensure as a Professional Engineer.

Ideally, engineering programs in are accredited by ABET every 6 years after a site visit by an ABET team. Typically, the program submits a self-study report few months prior to the site visit by the ABET team. The self-study can be considered a vital step for academic institutions to review the "health" of their engineering and technology programs and make improvements to enhance student learning. ABET expects every accredited program to collect assessment data on a regular basis and demonstrate continuous improvements in the snapshot of the self-study. Cummings et al. (2011) supported the process of self-critic before an ABET visit and wrote: "By providing a process to identify shortcomings with respect to student attainment of student learning outcomes, institutions or programs are better able to make necessary improvements, as well as understand their institutional or program strengths." Some programs have used the weaknesses or concerns raised by the ABET accreditation process to foster support from their administrations by requesting more financial support, more faculty and improvement of facilities and resources. Since their establishment in 1932, ABET standards have evolved over the years and have resulted in nationwide curricular revisions and the strengthening of engineering education (Shryrock et al., 2009). However, the potential for this system to drive improvement of engineering programs at academic institutions continues to depend on how well faculty understand, appreciate, and support the changes to meet these standards (Felder and Brent, 2003).

The first step in the accreditation process is to develop an assessment plan for an engineering program by involving faculty, students, and an Engineering Advisory Board that comprises external members from diverse backgrounds. This plan sets up program mission, program educational objectives, student outcomes, performance indicators (PIs), performance thresholds, and curriculum mapping. Establishing a data collection cycle and looping findings to make improvements are key components of the assessment process. A key step to appropriately assess student outcomes is to develop specific and measurable PIs that students are expected to demonstrate in relation to those outcomes. These PIs could then be incorporated into the course objectives. The PIs can be mapped to appropriate courses and measured using the appropriate assessment tools. Selecting appropriate academic tools to adequately assess student performance on the outcomes is crucial since focusing on overall course grades does not provide important details on student learning. Performance threshold(s) for meeting the individual outcomes must be established before the assessment data are collected and reviewed.

The collective assessment of student learning and ABET Student Outcomes used in the Civil Engineering Program at United States Coast Guard Academy (USCGA) includes two key components:

1. *Assessment at the Course Level* – This is accomplished by conducting a review of all courses and how students performed in those courses at the end of each semester. This End of Course Review (EOCR) involves a meeting

of all faculty after final grades are submitted. The meeting is chaired by the Program's Assessment Coordinator and the Program Chair. During the meeting, the draft reports prepared by the course coordinators are discussed in relation to the assessment data collected in each course offered that semester. The assessment data for each course is presented for discussion by the course coordinator. This exchange helps faculty to discuss issues for continuous improvement and to plan future course offerings. This process ensures that faculty are accountable and consistently collect and summarize assessment data required for the continuous improvement of courses and the Program. Student success in meeting PIs is assessed and evaluated, and recommendations for course improvements are made. The product of these meetings is a final report for each course that documents key information on course management, assessment data, student performance, as well as recommendations for improvement. Results from the EOCR reports provide vital input for the Civil Engineering Program Review.

2. *Assessment at the Program Level* – This involves an evaluation of the program as whole. This is conducted every 2 years at the end of the spring semester. Two or more full days are reserved for all faculty to meet and participate in reviewing the previous two assessment cycles. The Program Chair and Assessment Coordinator prepare a detailed agenda of all the action items recommended in the previous Program Review along with summaries of assessment data. At this level, student performance on all ABET student outcomes for the entire program is evaluated holistically based on information from all the end-of-course review reports. Achievement of the program educational objectives is assessed and evaluated using several surveys and other tools (alumni, senior exit, supervisor/employer, capstone surveys, and interviews).

The various assessment activities undertaken throughout the academic year are shown in Figure 6.1. Only the EOCR and its relevance to the assessment of the skills and outcomes presented in the previous chapters are discussed further.

6.2   END OF COURSE REVIEW

The EOCR plays a major role in the process used to assess individual student outcomes and to evaluate the effectiveness of the courses in preparing students to meet all the ABET Student Outcomes. In preparing the EOCR for each engineering course, student assessments are conducted throughout the semester using both direct and indirect measures. In general, the direct measures used in several courses are graded student work, such as homework, exams, projects, lab reports, technical reports, technical papers, and presentations. Several indirect measures in the form of surveys and questionnaires are completed by students, and other constituents for assessment of the course objectives, course outcomes, and the program educational objectives. In-class student and online surveys are conducted at the end of each semester in each course. In many courses, mid-semester in-class surveys are conducted as well. These surveys include general feedback on the course and may

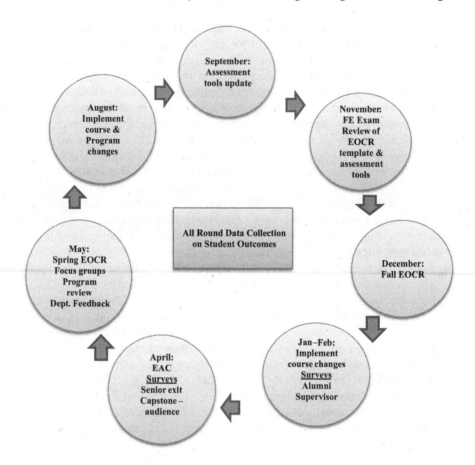

FIGURE 6.1 Annual assessment activities cycle.

also include student self-assessment of their competency in the various content areas of the course. Prior to the EOCR meeting, each course coordinator prepares a draft report that includes a review of issues raised at the previous EOCR, an examination of assessment instruments used, the course's contribution to the program in the context of the Student Outcomes Achievement Matrix and course changes (long and short range). The final EOCR report contains the contents of the draft report, issues discussed at the meeting, action items, and a number of attachments. The final reports are stored electronically for future use and historical records. The detailed reflections contained in the EOCR documentation provide the continuity of information necessary to keep the course and curriculum focused on program, departmental, and institutional improvements.

In assessing student performance in various competencies and ABET student outcomes, the Civil Engineering faculty use several PIs to serve as measurable criteria students must meet for successful demonstration of achievement of the outcomes or development of an acceptable level of competency. The PIs serve as measurable "skills, knowledge and behaviors" students are expected to demonstrate regarding the outcomes or course objectives. These PIs are mapped to courses within the civil engineering curriculum. Assessment data are collected on the PIs to measure student achievement of the student outcomes. Most PIs have at least two courses providing direct assessment data; however, in a few cases, only one course is appropriate to provide direct assessment data. These primary courses represent courses that the faculty identified as providing the highest and most appropriate coverage of the PIs.

6.2.1 HOLISTIC ASSESSMENT DATA ANALYSIS

The assessment process should focus on student learning and the measurement of ABET student outcome achievement to support continuous improvement and is based on collaborative work among faculty members. The assessment process involves measurable PIs that are linked to each student outcome to allow for direct measures of outcome achievement using various assessment tools. In the Civil Engineering Program at USCGA, these tools include grades on student assignments or portions of assignments such as homework, projects, lab reports, technical papers, oral presentations, and exam questions. In many cases, rubrics were developed to facilitate the measurement of achievement on specific PIs. Thresholds and performance targets representing minimum student performance were established for the successful achievement of the PIs as shown in Figure 6.2. Different performance targets were set for exams and non-exam activities (projects, homework, reports, technical paper, oral presentations, etc.). A student is considered as having demonstrated satisfactory achievement of a PI if his/her score (grade on particular assessment tool) meets or exceeds the *PI Score* of

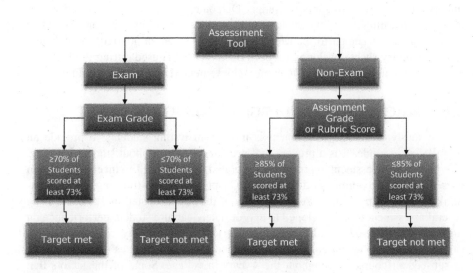

FIGURE 6.2 Assessment tools thresholds and performance targets.

73%. For a given course to be classified as producing satisfactory student achievement on a PI, one or both of the following performance targets must be met:

- *Exams* – at least 70% of students must exceed the *PI Score of 73%.*
- *Non-exams assignments* – at least 85% of students must exceed the *PI Score of 73%.*

The performance targets must be met for *all* the PIs in *each primary course* mapped to that PI for an outcome to be achieved.

Assessment data based on this approach are tabulated and documented in the EOCRs similar to the example shown in Table 6.1. When an assessment tool indicates a PI target is not met or there is a concern about student performance, the course coordinator and other faculty members offer suggestions for improvement during the EOCR to be implemented in future course offerings. The recommendations for improvements are documented in the final EOCR report to ensure that future course coordinators can follow through with the recommended changes.

The connection between EOCRs and student performance on the outcomes is made through a *Student Outcomes Achievement Matrix* that is used to holistically evaluate and track assessment data on the PIs linked to the student outcomes. This matrix shows the percentage of students meeting the PI score threshold (i.e., a grade of at least 73% on graded assignments). The *Student Outcomes Achievement Matrix* therefore provides a summary of performance on all the indicators for a holistic assessment of the Student Outcomes across the curriculum. This matrix is completed every academic year and informs a broader discussion on course and curricular improvements needed to support Student Outcome achievement and improvement of student learning. The Student Outcome Achievement Matrix is color coded to visually indicate when performance targets are met (green) or not (red). An example of the summary sheet developed for each outcome that is used during the Program Review to populate the Student Outcome Achievement Matrix is shown in Figure 6.3.

6.3 CRITICAL THINKING/DESIGN THINKING ASSESSMENT

As previously discussed, assessment is more than testing students to provide an overall course grade. It is a process that provides feedback about student learning. Therefore, the assessment approach incorporated by the Civil Engineering Program uses Bloom's Taxonomy. As discussed in Chapter 2, design-thinking and critical-thinking competencies are assessed in each of the courses listed in Table 2.4 using several assessment tools developed to appropriately capture student performance on the various cognitive levels. It is a challenging task to consistently track the progression of students in each course and the influence of prior courses on performance in subsequent courses throughout the 4 years in college. Some of the factors that make this challenging include student interest in the course material, differences in course requirements and types of assignments, and instructor teaching style, etc. At USCGA, more emphasis is placed on student performance during the senior year to assess design thinking and critical thinking competencies before graduation and transition into the civil engineering profession. The assessment tools have enabled

TABLE 6.1
Example of Student Outcome Assessment Summary Presented in EOCR Report (Geotechnical Engineering Design)

Outcome[a]	Performance Indicator[b]	Assessment Tools[c]	Average Score on Performance Indicators (High, Low Scores)[d]	% Students meeting Performance Indicator Score[e]	Performance Target for Performance Indicator Met?[f]
ABET 3-1: An ability to identify, formulate, and solve complex engineering problems by applying principles of engineering, science, and mathematics.	ABET 3-1-4: Apply theories, assumptions, & principles to the problem by demonstrating the use of relevant formulae & relationships; verify results.	Projects (weighted average)	80% (93.2%, 70.5%)	86.4%	Yes
ABET 3-2: An ability to apply engineering design to produce solutions that meet specified needs with consideration of public health, safety, and welfare, as well as global, cultural, social, environmental, and economic factors.	ABET 3-2-2: Apply engineering principles and design concepts (including design codes and specifications) to solve a civil engineering problem.	Projects (weighted average)	80% (93.2%, 70.5%)	86.4%	Yes
ABET 3-7: An ability to acquire and apply new knowledge as needed, using appropriate learning strategies.	ABET 3-7-4: Use modern engineering tools to foster critical thought in analysis and design.	Project 4	75.4% (90%, 62%)	63.6%	No

a List student outcomes assessed in this course.

b List the corresponding performance indicators related to the outcome in column 1.

c List assessment tools for which data was collected and reported in column 4. If multiple assessment tools are used, only list those that will support the results in columns 4–7.

d List average (could be weighted), high and low scores. Input numerical values in percentages. If rubric scores or rating such as 3, 2, 1 are used, convert to %.

e The passing grade for each performance indicator is set at a grade of (C or 73%). Calculate the percentage of students who scored at least 70% and report in column 5.

f Performance target for the PI is set at 70% for exams and 85% for non-exam assessment tools such as homework and project assignments, technical paper reports. The performance target is met if at least 70% of students score 73% or higher in exams and at least 85% of students score 73% or higher in non-exam assignments.

ABET 3-3: An ability to communicate effectively with a range of audiences.

PERFORMANCE INDICATORS

3-3-1: Use appropriate presentation tools and techniques to orally communicate information, concepts and technical ideas effectively.

3-3-2: Prepare written documents in standard engineering format to communicate information, concepts and technical ideas effectively.

3-3-3: Research information from a variety of sources, utilize information to make engineering decisions/judgment and produce a technically sound report.

3-3-4: Respond to questions from diverse audiences with justified and well formulated answers.

PRIMARY COURSES USED FOR ASSESSMENT

1304 Soil Mechanics

1309 Environmental Engineering I

1402 Civil Engineering Design

ASSESSMENT

Performance Indicator	Primary Courses	Semester	Course Coordinator	Assessment Tools	Results (% of students who scored ≥ 73%)	Achieved?*
3-3-1	1304	Spring		Technical paper oral presentation	100%	YES
	1402	Spring		Capstone final project presentation	100%	
3-3-2	1304	Spring		Technical paper	100%	YES
	1402	Spring		Capstone project final report	100%	
3-3-3	1309	Fall		Water Quality Lab report	100%	YES
	1402	Spring		Capstone project final report	100%	
3-3-4	1304	Spring		Technical paper oral presentation	100%	YES
	1402	Spring		Capstone final project presentation	100%	

*Yes, if results are >70% for exams and >85% for all other assessment tools.

SUMMARY OF OUTCOME SCORES

Performance targets were met on all four performance indicators. Therefore, student outcome "3-3" is achieved. Results from the 2019 senior exit survey indicate that 51.5%, 39.4% and 9.1% of students rated themselves as having full, moderate and fair competency related to this outcome, respectively. 100% of the respondents of the supervisors' survey rated students' demonstration of this outcome as excellent.

RECOMMENDATIONS

Even though all PIs were met, the final Capstone papers could have been much stronger. Future plans include increased number of submittals prior to the final submission. Continue asking students to provide customer brief. Earlier start to submissions. ASEE conference. It is anticipated that students will have increased opportunities to develop and hone their communication skill when the new core curriculum is implemented for the Class of 2021. The new core curriculum will include an information literacy thead that will be progressively and continuously assessed academy -wide in several courses. It is expected that this together with the harmonization of the labs and technical writing assignments across the CE curriculum will further strengthen student perfomance on this outcome.

FIGURE 6.3   Class of 2020: example of student outcome summary sheet for ABET 3-3.

civil engineering faculty to assess the competencies that students have gained in enhancing their critical thinking skills and in the design process.

Assessment in the first two upper-level (senior-year) courses shown in Table 2.4 is presented here as an example. A common grading rubric (Table 6.2 and Appendix 5) is used concurrently in the *Geotechnical Engineering Design* and *Reinforced Concrete Design* courses to assess students' critical thinking and design competencies as well as communication skills. Based on this rubric, the average performance of students for the graduating class of 2020 is shown in Figures 6.4 and 6.5 for the

TABLE 6.2

Common Project Grading Rubric

	Detailed Description	Exceeds Expectation (>90%)	Meets Expectation (70%–89%)	Below Expectation (<70%)
Define (Purpose of project) ___/5	Define the problem statement, identify constraints and scope of the project. *Bloom's Level 1*	Clear & concise general description of the project scope. Purpose very clearly stated; provides all key details as expressed in the project description. Demonstrates good understanding of problem/ assignment	Concise description of the project scope. Purpose clearly stated; excludes some one or more key details in the project description. Demonstrates fair understanding of problem/assignment.	No general description of the project scope. Purpose unclear; key details excluded; missing important information. Demonstrates poor or no understanding of problem/assignment.
Research (Investigate & gather relevant information) ___/5	Identify & explain objectives, gather information from relevant sources such as codes and specifications, and list assumptions. *Bloom's Levels 1, 2 & 3*	All project objectives are listed & clearly explained. Relevant sources of information properly cited. Demonstrates very good understanding of project scope. All the important assumptions are discussed.	Objectives are identified & listed, but not all are clearly explained. Relevant sources of information properly cited. Demonstrates good understanding of project scope. Most, but not all of the important assumptions are discussed.	Objectives are not identified & listed. Relevant sources of information not properly cited. Demonstrates poor understanding of project scope. Most of the important assumptions are not discussed.
Identify and Decide (Approach to developing & selecting design solutions) ___/10	Identify, select, & formulate appropriate analysis, design method(s), & design parameters. Consider practical constraints and constructability issues. *Bloom's Levels 1, 2, & 3*	Selected suitable analyses & formulated design approaches. Very good engineering judgment and inference; very good understanding of parameters & methods relevant to design.	Identified analyses & formulated design approaches that are mostly suitable (at least one not fully applicable). Good engineering judgment and inference; good understanding of parameters & methods relevant to design.	Suitable analysis & design method not identified or formulated. Poor engineering judgment and inference; little or no understanding of relevant parameters needed for design.

(Continued)

TABLE 6.2 (Continued)
Common Project Grading Rubric

	Detailed Description	Exceeds Expectation (>90%)	Meets Expectation (70%–89%)	Below Expectation (<70%)
Solve (Apply appropriate methods to analyze and design) ___/50	Conduct detailed analysis and document design calculations. Justify selected design approach. *Bloom's Levels 3, 4, and 5*	Fully applied principles & relevant code with no conceptual or computational errors. Correct calculations & estimation of design parameters used appropriate correlations. Design is correct & complete. All constraints adequately addressed. Very clear justification of design approach.	Applied principles & relevant code. Few (<5) conceptual or computational errors. Correct estimation of design parameters used appropriate correlations, accuracy adequate. Most of the key constraints are addressed.	Incorrect calculation of design parameters used inappropriate correlations. Inability to apply principles & relevant code. Incomplete and/or inaccurate design. Most of the constraints are not addressed.
Verify (Evaluate Solutions, make Conclusions & recommendations) ___/10	Conduct analysis. Ensure the proposed design adequately addresses all aspects of the problem statement and scope of the project. Exercise judgment & check if design is practical & reasonable. Make recommendations to address pending issues. *Bloom's Levels 5 & 6*	Very good engineering judgment on adequacy of proposed design & decisions are fully supported by code & design requirements. Degree to which solution meets project objectives is well explained. Recommendations address all issues of practicality & constructability.	Good engineering judgment on adequacy of proposed design & decisions are supported by code/design requirements. Degree to which solution meets project objectives is fairly explained. Recommendations address some but not all issues of practicality & constructability.	Poor engineering judgment on adequacy of proposed design & decisions are not supported by code/design requirements. Degree to which solution meets project objectives is not explained. Recommendations address some but not all issues of practicality & constructability.
Design Drawings ___/10	Illustration design results/recommendations. Drawings & sketches showing relevant dimensions & important details ***drawn to scale*** in accordance with design calculations. *Bloom's Levels 2 & 3*	Drawings & sketches are complete, accurate in accordance with design calculations, & drawn to scale.	Drawings & sketches are drawn to scale in accordance with design calculations, accurate but 10%–20% incomplete.	No drawings or sketches or drawings are inaccurate or drawings not based on design solution or >40% incomplete,
Report Organization ___/10	Professionalism & quality of report. *Bloom's Levels 2 & 3*	Project report is logically organized, well developed, very clear & easy to follow calculations, good transition, good formatting; no grammatical errors.	Project report is logical with good transition and flow, calculations easy to follow, no Few (<5) grammatical errors; little (<3) formatting errors.	Project report is unorganized, poor transition, calculations difficult to follow, several grammatical and formatting errors.

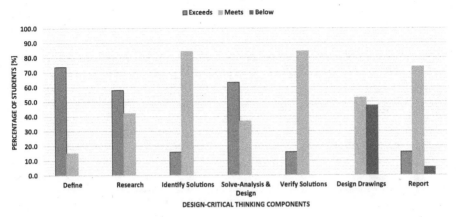

FIGURE 6.4　Class of 2020: assessment of student performance in geotechnical engineering.

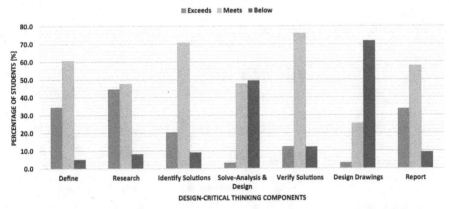

FIGURE 6.5　Class of 2020: assessment of student performance in reinforced concrete design.

Geotechnical Engineering Design and Reinforced Concrete Design course, respectively. These figures show the average performance on all the projects at the end of the semester. The various components of the rubric include critical thinking aspects that are based on Bloom's levels. By capturing student performance on these components, their critical thinking development and design skills can be progressively tracked as they complete the assignments. This assessment method places students into one of three categories (exceed, meets, or below expectation).

Figures 6.4 and 6.5 indicate a significant percentage of seniors underperformed in the "drawing documentation" criterion in both courses. Developing design drawings

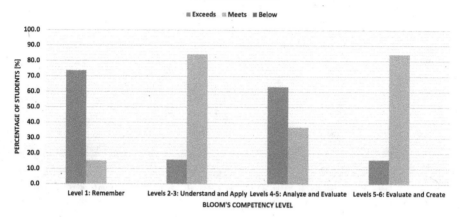

FIGURE 6.6 Class of 2020: students bloom's cognitive competencies in geotechnical engineering course.

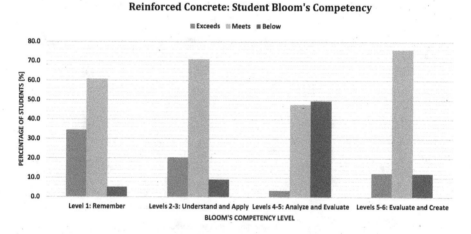

FIGURE 6.7 Class of 2020: students bloom's cognitive competencies in reinforced concrete course.

and documentation was identified as an area for improvement in future course offerings. Part of the reason why student underperformed in this area is that there is no drafting or engineering drawing requirement in the curriculum. Therefore, some students struggle to visually represent details of their project calculations and solutions. About 50% of students performed "below expectation," in the reinforced concrete course in the "solve-analysis & design" criteria mainly due to the open-endedness of the project assignments. The open-ended structure of assignments challenged students to think more deeply about the solutions and cognitive process engaged students at high levels of Bloom's taxonomy. The data for the same cohorts (Class of 2020) in terms of Bloom's cognitive levels in both courses are shown in Figures 6.6

TABLE 6.3

Class of 2020: Critical Thinking Competency Survey Results for Geotechnical Engineering Course

2001 Bloom's Taxonomy Critical Thinking Domain	Description (Action Verbs)	% of Students Indicating Competency
Remember/Knowledge	I can recall facts and basic concepts (define, duplicate, memorize)	87%
Understand/Comprehension	I can explain ideas and concepts (classify, describe, identify, discuss)	78%
Apply/Application	I can use information in new situations (implement, execute, solve, sketch)	78%
Analyze/Analysis	I can draw connections among ideas (differentiate, organize, compare, examine)	83%
Evaluate/Evaluation	I can justify a stand or decision (defend, appraise, select, support)	83%
Create/Synthesis	I can produce new or original work (design, assemble, construct, develop, investigate)	65%

and 6.7. All indications show that these exercises provide students with opportunities to reach the higher levels of the Bloom's Taxonomy.

Furthermore, an indirect assessment in the form of a critical thinking self-assessment survey was administered at the end of the semester in the Geotechnical Engineering course for the Class of 2020 (Appendix 6). The survey required students to indicate their level of cognitive competency achieved in the course. The results summarized in Table 6.3 show about 80% or more of the students indicated competency in levels 1–5 of Bloom's Taxonomy and 65% indicated competency at the highest level (6) of Bloom's Taxonomy. The self-assessment results are somewhat in agreement with the graded assignments except for level 6. Nonetheless, the overall assessment data indicate that students are making progressive improvement in their problem-solving abilities, performing better on their senior capstone design projects, and showing better preparedness to make the transition to practice engineering after graduation.

6.4 INFORMATION LITERACY AND COMMUNICATION SKILLS ASSESSMENT

Information technology continues to advance and the need for effective communication and information literacy skills in engineering practice were discussed in Chapter 4. Examples of the assessment data collection and analysis in the junior and senior years are presented in this section. The technical paper assignment in the junior year that is used to foster the development of information literacy skills is shown in Figure 6.8, and the grading rubrics in Tables 6.4 and 6.5.

The assessment data on student performance based on the components of the rubrics are shown in Figures 6.9 and 6.10. This aggregation of the data enables

Soil Mechanics Technical Paper Requirements

The technical paper represents 13% of the course grade. It is assigned to strengthen three very important skills; conducting a literature search, writing a technical paper, and public speaking. Your technical research paper should demonstrate that you are developing the necessary skills to become an independent, life-long learner. An assessment rubric will be used to grade your paper and assess your competencies in the following areas:

1. **Information Literacy**: Information literacy is defined by the National Forum on Information Literacy as; *"The ability to know when there is a need for information, to be able to identify, locate, evaluate, and effectively and responsibly use and share that information for the problem at hand."* This includes demonstrating why your topic is important in the context of soil Mechanics and Foundations, and to be able to identify the type of information and extent of research that is appropriate to support your topic. Information literacy also includes an ability to search for and use a variety of valid technical information sources, to quote and paraphrase sources appropriately, and to use proper methods of citation.

2. **Technical (Content) Competency:** The ability to express technical material in a clear and grammatical manner at a level appropriate to your audience.

3. **Writing Skills:** The ability to produce a well-organized and professional research paper as per stated guidelines or format that makes use of appropriate sub-sections and properly incorporates Tables and Figures into the text.

Organization:

The exact sections and sub-sections used will depend on the nature of each topic. At a minimum the paper should contain the following sections:

a.   Abstract

b.   Introduction – The introduction should start with a general description of the topic.

c.   The main research section – This section **should be broken down into logical sub-sections** that flow naturally from a detailed outline you will develop before writing your paper.

d.   The conclusion – This section should tie together the main concepts from the paper and relate them back to the overall topic. As appropriate, suggestions for future research in your topic can be made.

e.   A bibliography that lists all references used in ASCE format or similar format.

The length of the paper should not exceed 10 pages without the bibliography. Additional details on the organization and format of the paper are provided on a separate handout. A paper may be longer if your topic warrants the extra length, however, extra length caused by poor writing or poor organization will cost you points.

Referencing:

All sources must be properly cited in the text and must be listed in a bibliography at the end of the paper using the ASCE journal guidelines (available at http://www.asce.org) or the guidelines distributed in class. All tables and figures must be numbered sequentially, given a meaningful title, introduced in the text by number (i.e. Figure 1 shows ...), and should be incorporated into the body of the paper as soon as possible after being introduced. **A minimum of 4-6 appropriate technical references** should be used from a variety of sources such as technical books, manuals, reports, journal articles, web sites, etc.

FIGURE 6.8   Class of 2020 soil mechanics technical paper requirements (junior year).

faculty to easily identify areas where students struggle or underperform and to plan accordingly.

By their senior year, students would have had several opportunities to hone their general communication and information literacy skills. The culminating test of these skills takes place in the Civil Engineering Design (CED) capstone design course. Performance on the CED final report for the graduating class of 2020 is shown in Figure 6.11. At least 80% of the students either met or exceeded expectations in each component of the CED grading rubric. It is important to note that faculty members typically have higher expectations for seniors and, in some cases, the actual achievement of PIs may appear to be the same or higher in the junior level classes. This phenomenon is discussed and taken into consideration during biennial program review meetings and also at end of course review meetings.

6.5   LEADERSHIP SKILLS ASSESSMENT

The assessment of leadership skills development over the 4 years is a continuous process that starts during the summer training camp (swab summer). During swab summer students are introduced to the various leadership styles and structure in the military. During the freshman year, students learn how to lead themselves and they

TABLE 6.4

Class of 2020 Soil Mechanics Written Paper Grading Rubric

Assessment Areas	Exceeds Expectations (90%–100% of Allotted Points)	Meets Expectations (70%–89% of Allotted Points)	Below Expectations (<70% of Allotted Points)
Ability to identify the type and extent of information needed. ____/10	The introduction clearly articulates the relevance of the topic to soil mechanics and/or foundation engineering. The scope and extent of the research in the entire paper provides excellent background to support the topic.	The introduction touches on the relevance to soil mechanics and/or foundations. The scope and extent of research in the entire paper provides some good background to support the topic but is not comprehensive.	There is little to no mention of the tie between the topic and soil mechanics and/or foundations. The type and extent of research is inadequate to provide background for the topic.
Ability to search for and incorporate a variety of appropriate technical information sources. ____/20	At least six appropriate (and credible) technical information sources are used that represent a variety of sources (i.e., technical books, reports, manuals, web sites, journal articles, etc.). The sources are varied and comprehensive enough to fulfill the scope of the research without over reliance on one or two sources.	At least four appropriate (and credible) technical information sources are used that represent at least type types of information sources. The sources generally support the scope of the research and no one source is too heavily relied upon.	There are less than four technical information sources used and/or one or more sources are inappropriate for a technical research paper. One or two sources are too heavily relied upon. The sources do not adequately cover the scope of the research. One or more source not credible.
Ability to use sources appropriately, legally, and ethically. ____/10	Sources are properly paraphrased, and all direct quotes are in quotation marks. References are cited properly within the text and are appropriately listed in a bibliography at the end of the paper according to the ASCE referencing guidelines.	Sources are properly paraphrased and direct quotes are in quotation marks except for minor issues. References are in the bibliography using the ASCE or similar referencing guidelines.	Sources are not properly paraphrased and/or direct quotes are not properly attributed. Citations are missing from the text and/or are not properly listed in the bibliography according to the ASCE or similar referencing guidelines.

(Continued)

TABLE 6.4 (Continued)
Class of 2020 Soil Mechanics Written Paper Grading Rubric

Assessment Areas	Exceeds Expectations (90%–100% of Allotted Points)	Meets Expectations (70%–89% of Allotted Points)	Below Expectations (<70% of Allotted Points)
Technical content ___/40	All technical aspects are covered with sufficient details in the context of topic. All important conclusions are clearly stated. Student shows excellent understanding of the relevance of the topic to soil mechanics and foundation engineering.	Most of the important technical details are covered. Only minor details missing. Conclusions regarding major points are stated. Student shows good understanding of the relevance of the topic to soil mechanics and foundation engineering.	Several important technical details are missing or not addressed in sufficient details. Conclusions are missing important aspects of the topic. Student shows lack of understanding of the relevance of the topic to soil mechanics and foundation engineering.
Technical writing skills. ___/10	Writing is clear, concise, and easy to understand. Sentences are grammatically constructed. Passive voice is used. There are no typos or spelling errors. Tables and figures are numbered and properly incorporated into the text after being introduced.	Writing is fairly easy to understand with a few unclear or poorly constructed sentences. There are a few typos or spelling errors. Tables and figures are properly used and incorporated.	Writing is of poor quality. The concepts are poorly explained. There are multiple grammatical errors and/or typos. Tables and figures are not numbered, properly incorporated into the text, and/or introduced.
Ability to format a well-organized paper. ___/10	The paper is well organized as per guidelines or format provided. The introduction explains how the research supports the topic. Subheadings flow logically. The conclusion ties together the main ideas.	The paper is fairly well organized with an introduction, main research section with subheadings, and a conclusion. The conclusion ties together the main ideas. Formatting generally good as per guidelines provided.	The paper lacks good organization. The paper is missing an introduction or conclusion. Formatting does not follow guidelines provided.

TABLE 6.5

Class of 2020 Soil Mechanics Oral Presentation Grading Rubric

Assessment Areas	Exceeds Expectations (90%–100% of allotted points)	Meets Expectations (70%–89% of allotted points)	Below Expectations (<70% of allotted points)
Organization ___/20	Very clear introduction, conclusion and consistent sequenced material within body of the presentation. • Information in logical, *interesting* sequence that audience can *easily* follow. • *Engaging* beginning and/or thoughtful ending. Moves smoothly from one idea to the next all of the time.	Clear introduction, conclusion and consistent sequenced material within body of the presentation. • Information in logical sequence that audience can follow. • Appropriate but not engaging beginning or ending. Moves smoothly from one idea to the next some of the time.	No introduction and/or conclusion and unsequenced material within body of the presentation. • Sequencing is difficult to follow. • Lacks beginning or ending, or inappropriate beginning or ending. Does not move smoothly from one idea to the next.
Technical Content (Coverage of subject matter) ___/40	A variety of supporting materials (e.g., illustrations, statistics, analogies, quotations) make appropriate reference to information or analysis that *significantly* supports the presentation. • Topic covered with sufficient details. • Student demonstrates full knowledge. • Can answer questions fully and accurately.	Supporting materials (e.g., illustrations, statistics, analogies, quotations) make appropriate reference to information or analysis that *generally* supports the presentation. • Most important aspect covered but some minor details missing. • Student demonstrates knowledge of basic concept. • Can answer questions but not fully.	Insufficient supporting materials (e.g., illustrations, statistics, analogies, quotations) make appropriate reference to information or analysis that minimally supports the presentation. • Topic covered with insufficient details. • Student lacks understanding of information. • Cannot answer questions.
Technical Accuracy of Content ___/15	Information is correct and accurate.	Some errors that could distract a knowledgeable listener, but most information accurate.	Information is inaccurate to the extent that presentation cannot be seen as a source of accurate information.

(Continued)

TABLE 6.5 (Continued)
Class of 2020 Soil Mechanics Oral Presentation Grading Rubric

Assessment Areas	Exceeds Expectations (90%–100% of allotted points)	Meets Expectations (70%–89% of allotted points)	Below Expectations (<70% of allotted points)
Style and Delivery ___/20	Delivery techniques (posture, gesture, eye contact, and vocal expressiveness) make the presentation *compelling*, and speaker appears polished and confident. • Listeners are captivated & focused on ideas presented. • Referred to notes or slides but did not read from them. • Relaxed body language. • Consistent eye contact. • Voice is clear with interesting modulation.	Delivery techniques (posture, gesture, eye contact, and vocal expressiveness) make the presentation *understandable*, and speaker appears tentative. • Listeners can follow presentation, but some distractions. • Reads from notes or slides but also looked away from them. • Slight nervousness. • Occasional unsustained eye contact. • Voice with some inflection.	Delivery techniques (posture, gesture, eye contact, and vocal expressiveness) detract from the understandability of the presentation, and speaker appears uncomfortable. • Listeners are distracted, with difficulty to follow. • Reads from notes or slides & did not look away from them. • Nervous body language. • No effort to make eye contact. • Monotone voice.
Format ___/5	Uniform and consistent fonts and colors. Background and layout of information are appropriate.	Background is appropriate. Good layout, but minor inconsistencies with fonts and colors.	Poor background and inconsistent fonts. Colors, background, and layout inappropriate & very distracting.
Timeline (coordinator's responsibility) [Points Deduction]	Presentation was within 2 minutes of allotted time. [*No deductions*]	Presentation was within 2–5 minutes of allotted time. [−1 point deduction/minute]	Presentation was 5 minutes over or under the allotted time. [−2 points deduction/minute]

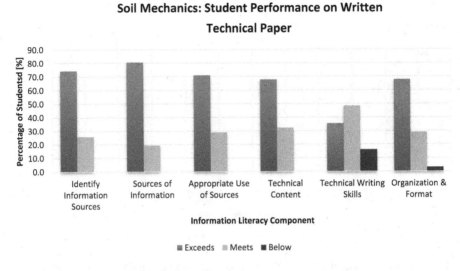

FIGURE 6.9 Class of 2020: information literacy skill competency (junior year).

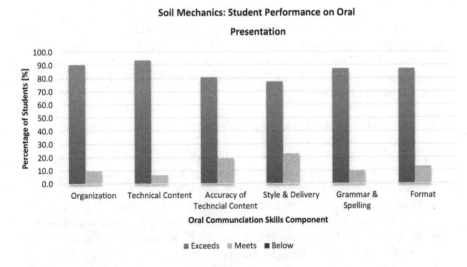

FIGURE 6.10 Class of 2020: oral communication skills competency (junior year).

are continuously evaluated by their mentors or supervisors both formally and informally. With each progressive year, students shift their focus to the next set of competencies in the leadership development framework discussed in Chapter 3. Within the military setting, continuous feedback on leadership development and the associated competencies is provided using the "Cadet Evaluation Report" (CER). Sophomores have the opportunity to transition into campus and Corp-wide leadership positions where they are expected to practice leading others and by being role models.

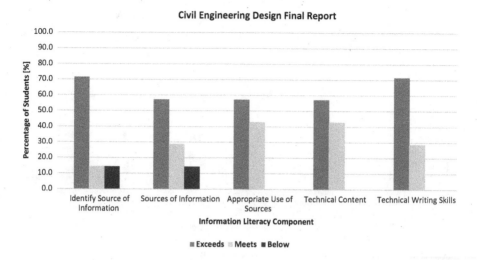

FIGURE 6.11   Class of 2020: communication and information literacy skills competency (senior year).

Within the academic setting, sophomores are required to take the *Leadership and Organizational Behavior* course where students are exposed to fundamental leadership and management concepts as discussed in Chapter 3. The focus of this course is on leading self and leading others thus providing a basis for students to understand the fundaments of leadership that they can then apply in other settings. The course objectives include the following:

- Be able to understand your personal leadership style and the strengths and weaknesses of that style
- Be capable of entering into a team environment in a leadership role, assess the dynamics of that team, and adapt your leadership style to effectively lead the individual members and team as a whole
- Possess a thorough understanding of quality "followership" and be capable of developing high quality, more engaged, more critically thinking followers
- Be able to articulate, evaluate, and implement common academic and Coast Guard organizational behavior and leadership frameworks

A detailed description of the leadership development in the civil engineering curriculum was discussed in Chapter 3. A significant part of this development occurs outside the civil engineering department. However, there is a component addressed by faculty in the senior year of the curriculum. Seniors are required to complete two leadership assignments as part of the capstone course. A sample of one of the assignments is shown in Figure 6.12. The extent of the leadership assessment done by the civil engineering faculty is somewhat qualitative as students reflect on their experiences. The importance for students to learn and grow from their experiences is the focus of the academic leadership assignment shown in Figure 6.12.

Leadership Essay Assignment

Students will write two leadership journal entries throughout the semester. The objective of the assignment is to reflect on how leadership concepts apply to the project and team experience. Because of the nature of this assignment, outside sources are not required. However, students are encouraged to make use of all available information, including work done as a part of your Leadership Development in various academic courses, summer training programs, and general military training.

Essay Expectations:

Your first essay should, at a minimum, include:

- o A description of how you plan to approach your project from a leadership perspective;
- o An analysis of how this plan may differ from a more traditional leadership role due to th e nature of the project/nature of the group you are working in;
- o Explanation of at least two leadership challenges you anticipate facing during the project, and how you intend to overcome those challenges;
- o What you intend to learn from this experience (in terms of leadership).

Your second essay should, at a minimum, include:

- o A brief summary of your original journal;
- o Leadership lessons learned from the project;
- o A reflection on how you can make use of this experience in future situations.

Timeline:

- o Week 4 – Essay #1 due
- o Week 12 - Essay #2 due

Grading Criteria: You will be graded on the following:

- o Quality of writing
- o Depth of insight/reflection
- o Meets criteria set forth in assignment description
- o Grammar
- o Organization of entry

Format:
Written work should be 2-3 pages in length, double-spaced, Times or Times New Roman 12-point font.

Referencing:
All sources must be properly cited in the text and must be listed at the end of the journal using the ASCE journal guidelines posted on D2L and available at:
https://ascelibrary.org/doi/pdf/10.1061/9780784479018.ch05

Submission:
Submit essay with cover page in class at 0800 on assigned due date.

Collaboration:
Students may only consult with the course Instructor, any other Faculty member, or cadet Writing & Reading Center Staff when researching and writing Journal Assignments.

FIGURE 6.12   Class of 2020: leadership assignment-senior year.

REFERENCES

ABET. (2018). *Criteria for Accrediting Engineering Programs.* ABET Engineering Accreditation Commission, Baltimore, MD.

Cumming, T., Heng, I. and Tsang, R. (2011). "Using direct assessment to resolve TAC/ABET criterion 3 program outcomes." *Proceeding of the Annual American Society of Engineering Education Conference*, Vancouver.

Felder, R. and Brent, R. (2003). "Designing and teaching courses to satisfy the ABET engineering criteria." *Journal of Engineering Education*, Vol. 92, No. 1, pp. 7–25.

Shryrock, K. and Reed, H. (2009). "ABET accreditation: best practices for assessment." *Proceeding of the Annual American Society of Engineering Education Conference.* ASEE Annual Conference and Exposition, Austin, Texas. pp. 14.148.1 – 14.148.11.

Spurlin, J. E., Rajala, S.A. and Lavelle, J.P. (2008). *Designing Better Engineering Education Through Assessment.* Stylus Publishing Co, Sterling, VA.

7 Engineering Outreach
Recruitment and Retention

7.1 INTRODUCTION

It is well documented that the quality and quantity of engineers needed in the United States' engineering industry are not being adequately met by the current educational system. Although some of the nation's post-secondary institutions are experiencing growth in individual engineering programs, an overall general disinterest or indifference in engineering as a profession continues to capture the attention of admissions officers, college deans, and others who are concerned about the future of the engineering profession. Trends show that even though there has been rapid expansion in technological advancements and a growing demand for engineers in the USA, student graduation rates indicate that engineering schools have been unable to supply neither the number nor the quality of engineers required by industry. Due to the current globalization trends and increased outsourcing of jobs, engineers not only need to have analytical skills but must also be able to function in multicultural environments; possess good leadership, management, and interdisciplinary skills; and understand the need to be engaged in life-long learning. Additionally, there is a growing global need for engineers to possess an appropriate level of professional skills as well as public-policy awareness and cultural sensitivity.

The declining interest of high school and first year college students in certain engineering disciplines continues to be a concern for engineering departments nationwide in the United States. Several recruiting programs have been developed to promote the engineering profession as well as address the issue of declining interest and inadequate retention; to a large extent, most of these programs are short-term based. Most of these programs target high school students; are only applicable in one grade level; or are one-time project-based events. According to the "Science and Engineering Indicator 2018" report, from 2000 to 2009, there was an average annual increase of 1.8% in the number of bachelor's degrees awarded in engineering, and from 2010 to 2015, the average annual increase was 6% (National Science Board, 2018). This positive change could generally be attributed to the increasing size of the college age cohort and not to a rising interest among students to attend college or select a Science, Technology, Engineering, and Mathematics (STEM) major. Despite this increase, there is still a high attrition rate (50%) in many engineering programs. Unfortunately, some of the students who are now choosing non-engineering majors are among those who possess the strongest math and science backgrounds.

Why are students not selecting or leaving engineering? The reasons could include poor teaching performance of the faculty, difficult curriculum, poor student performance, especially in math, restrictive coursework, misperceptions of engineering, poor connection with the engineering community, isolation, and the need for

DOI: 10.1201/9781003280057-7

educational reform. The National Science Foundation reported that the science and engineering workforce is made up of only 15% of women and 10.5% of minorities (National Science Board, 2016). The same report indicated that:

- "Women have earned about 57% of all bachelor's degrees and about half of all Science and Engineering (S&E) bachelor's degrees since the late 1990s. Men earn the majority of bachelor's degrees in engineering, computer sciences, mathematics, statistics, and physics. However, women earn the majority in the biological, agricultural, social sciences and psychology."
- "Between 2000 and 2013, the proportion of S&E bachelor's degrees relative to degrees in all fields awarded to women remained flat. During this period, it declined in computer sciences, mathematics, physics, engineering, and economics."

Therefore, there has been a general decline in the number of females and minorities entering into engineering in spite of aggressive recruitment efforts by most post-secondary institutions. In the National Academy of Science 2018 report, there are still indications that undergraduate students who are capable of succeeding in Science, Technology, Engineering, and Mathematics (STEM) programs are still dropping out and switching to other career opportunities (National Academy of Science & Engineering, 2018). This poor retention rate was attributed to documented weakness in STEM teaching, learning, and student support. One of the conclusions in this report is in order to improve the quality and impact of undergraduate STEM education, the following areas must be adequately addressed: (1) students' mastery of STEM concepts; (2) justice, equity, diversity, and inclusion; and (3) completion of STEM credentials. Students' mastery can be increased by engaging them in evidence-based educational practices and programs. Institutions should also strive for equity, diversity, and inclusion so that students and instructors are provided with equitable opportunities for access and success. There should also be a push to ensure that adequate number of professionals complete their credentials in the various disciplines through the promotion and support for continuous improvement.

Several programs, sponsored by the National Science Foundation (NSF), American Society for Engineering Education (ASEE), National Society of Professional Engineers (NSPE), universities, and other institutions, have been implemented in an attempt to address the challenges associated with recruitment and retention of prospective engineers, especially in such underrepresented groups as women and minorities (NASA, 2016). The targeted groups have varied depending on the program sponsors' goals and objectives, and at times there have been some overlaps. Focusing on the improvement of math, science, and technology skills was the thrust of the vast majority of such programs, which were primarily designed to reach students at the high school and collegiate levels. Such programs have oftentimes enticed participants to become involved by offering opportunities for them to engage in real-world, hands-on engineering projects. The NSF has played a key role in funding such programs – many of which have been implemented by ASEE, NSPE, Kindergarten to 12th grade (K-12) schools, colleges, universities, and community action groups. As a result, efforts to develop curriculum, establish engineering education coalitions, promote research, and encourage the use of alternative pedagogies in an effort to enhance teaching and learning in STEM areas, including engineering, have benefitted from the availability of NSF grant opportunities. This resulted in several robotic competition programs currently promoted by a variety of organizations at the regional and national levels.

Although a considerable number of these programs have generally been viewed as effective and helpful, they have not resulted in long-term systematic changes in the perception and retention of engineers. What additional steps must then be taken to ensure that there will be enough engineers for the next century? Over two decades ago, Samuel C. Florman in an article published in the Technology Review stated, "Engineers should cultivate public interest in their profession by teaching elementary students the engineering process and promote the understanding of engineering concepts. Only then, can engineers hope to gain public appreciation of their contributions to society without having their innovations portrayed as exciting adventures in the media" (Florman, 1996). Engineering has been infused into the K-12 curricula in various ways, but the exposure compared with other subjects remains relatively small. Some progress has been made in introducing engineering in K-12; however, comments from an article published in 2009 are still valid:

> ...in contrast to science, mathematics, and even technology education, all of which have established learning standards and a long history in the K–12 curriculum, the teaching of engineering in elementary and secondary schools is still very much a work in progress

> *Katehi et al. (2009).*

There are well-established standards for science, technology, and math in K-12 education. In 2010, there was still no formalized standard for K-12 engineering education as reported in the National Academies Standard for K-12 Engineering Education report (National Academy of Engineering, 2010). It was concluded that

> although it is theoretically possible to develop standards for K–12 engineering education, it would be extremely difficult to ensure their usefulness and effective implementation. This conclusion is supported by the following findings: (1) there is relatively limited experience with K–12 engineering education in U.S. elementary and secondary schools, (2) there is not at present (2010) a critical mass of teachers qualified to deliver engineering instruction, (3) evidence regarding the impact of standards-based educational reforms on student learning in other subjects, such as mathematics and science, is inconclusive, and (4) there are significant barriers to introducing stand-alone standards for an entirely new content area in a curriculum already burdened with learning goals in more established domains of study.

To date, some of these challenges are yet to be overcome, although some progress has been made since the introduction of the "Next Generation Science Standards" in 2013 that promotes the engineering aspects of K-12 STEM education by connecting science concepts and practice to engineering (NGSS, 2013).

Additionally, there may be another reason for the slow-down in the number of students choosing engineering as a career: students all too frequently simply do not know what engineers do or how engineering affects the world. Results from a survey of elementary school students in the United States indicated that 85% of those

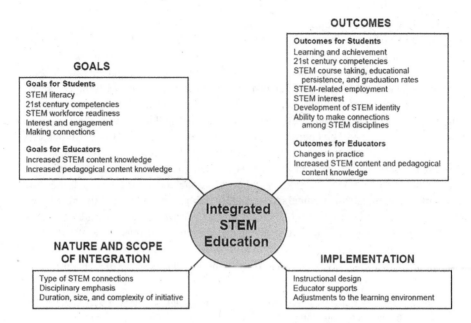

FIGURE 7.1   Descriptive framework for K-12 integrated STEM education (NRC, 2014)

surveyed were not interested in an engineering career. Of those expressing non-interest, 44% did not know much about engineering (ASCE News, 2009). Academic institutions and professional organizations such as ASEE have been leading the effort to promote engineering education in K-12. In this effort, ASEE has developed a set of standards to formally address the key aspects of the fundamentals of engineering in K-12 and professional development for teachers. These standards address five components important to the professional development of teachers: engineering content and practices; pedagogical content and knowledge; engineering as a context for teaching and learning; curriculum and assessment; and alignment to research, standards, and educational practices (ASEE P12 website). In a 2014 National Academies of Science report, a framework was developed that consists of four main features for an integrated STEM education in K-12 curriculum. Each of the four features includes the subcomponents shown in Figure 7.1 (NRC, 2014). It should be noted that engineering has the smallest footprint in K-12 education.

This chapter includes a general discussion of outreach programs, recruitment, and retention. The concept of "career imprinting" is introduced as an option to promote engineering in K-12 education as well as get students interested in an engineering career. Details of this approach are presented including examples of how engineering principles can be infused into the current K-12 curriculum. Potential challenges to implementation of such a concept are also discussed.

7.2   CAREER IMPRINTING

While there are numerous ways to encourage K-12 students to develop interest in engineering as a profession, a systemic change will only occur with a high rate of success when a well-developed strategy, that includes a systematic long-term grade level appropriate infusion of engineering concepts and ideas into all subject-matter disciplines of K-12 curricula. One such strategy is based on the principle of "career imprinting" to create interest in engineering as a profession. The concept of "career imprinting," similar to that espoused by Harvard Professor Monica C. Higgins in the field of business, is proposed as a possible strategy to increasing students' awareness of and appreciation for careers in engineering (Higgins & Schein, 2005). Through a series of shared classroom experiences, students in grades K-12 can be systematically introduced to engineering via the subject matter to which they are already being exposed, thus creating linkages/connections for students among their course content, learning environments, and personal daily lives that will "imprint" them to develop a strong career interest in a variety of engineering disciplines. The objective would be to not only assist students in cultivating and acquiring grade level appropriate, as well as discipline-specific basic skills needed by engineers, but also to help them develop the personal characteristics that are essential to be successful in the engineering profession. By connecting "engineering as a profession" to the development of these skills and characteristics, students may be sparked to have an interest in engineering and can become "imprinted" to consider careers in the field.

To cultivate students' interests in engineering as a profession, students must first somehow be connected to engineering in ways that entice or inspire rather than coerce them. Through the daily classroom instruction to which students are exposed, students can be imprinted to choose careers in engineering. Such infusion must start as early as possible in the educational development of students and be consistently reinforced throughout each grade level, if a long-term solution to the current declining interest in engineering is to be found. Short-term programs have their place in attracting students, but systematic and consistent exposure to the various components of engineering should be infused into current K-12 curricula if a significant increase in the number of future engineers is to be achieved. "Career imprinting" may, therefore, be a possible strategy for increasing the flow of future students into the engineering pipeline. Career imprinting is proposed as a strategy to foster student interest in engineering and help students develop confidence in their abilities to pursue any engineering career. Overall, this could be a good tool to educate K-12 grade students about engineering.

Career imprinting, as defined by Harvard Business School Professor Monica C. Higgins, is "the set of capabilities and connections, coupled with the confidence and cognition that a group of individuals share as a result of their career experiences at a common employer during a particular period in time" (Higgins & Schein, 2005). Applying this concept to education will enhance the "classroom experiences" in which Prekindergarten to 12th grade students engage at each grade level serve to imprint them. Thus, imprinting students to become interested in engineering necessarily involves not only introducing them to actual engineers but also connecting the subject matter they are taught to real-world applications in the field of engineering. Therefore, students could be "imprinted" to select careers in engineering if the relevance of engineering to their daily lives is made apparent. This can be achieved through teachers' conscientious and calculated creation of linkages among engineering, course content, and students' "real-world" experiences.

The "classroom experiences" students have during "school at each grade level" serve to imprint them. In this case, these experiences are referred to as "Engineering career imprints." The objective of the strategy is to help students develop the capabilities to make the right connections and build the confidence and cognition to become engineers through a series of shared experiences by introducing them to engineering via the subject matter to which they are exposed. The desired outcome is that the "imprints" made through these exposures will create linkages for students between engineering and course content, as well as among students' learning environments, personal daily lives, and the various engineering disciplines. This may then lead students to select careers in engineering.

Curriculum development initiatives such as *Engineering Is Elementary*: *Engineering and Technology Lessons for Children* (EiE) "capitalizes on youngsters' natural fascination with building, taking things apart, and figuring out how things work" (Museum of Boston, 2007, www.eie.org) by integrating elementary science and engineering concepts into separate activity-based lessons. The proposed career imprinting strategy, however, focuses on highlighting for students – through regular curriculum content – existing connections to engineering. As previously mentioned, compared to 20–30 years ago, technology now plays a more significant role in day-to-day living, and youngsters seem to naturally develop an aptitude for electronics/technology. Why is this the case today? Some would argue that at an early age, today's youth see the need and relevance, for example, operating a tablet, cellphone, video game, or other electronic devices. Youngsters, therefore, identify with and embrace the idea that technology is a necessary part of their daily lives. Thus, without much thought or even detailed instructions in some cases, they seem to be able to master a wide range of technology-based and relevant operations. The desire to be able to play a video game, use an App, participate in a robotics competition, operate a drone, and use a 3D printer instills a precise or particular imprint that causes a young person to be motivated to learn. The capacity for K-12 grade students to apply complex principles at an early age should be harnessed. This is an opportunity to introduce engineering ideas in elementary through high schools – with a focus on making it relevant to everyday living. This approach will encourage students to embrace it as a "way of life."

Teachers should be supported to expose students consistently and systematically to technical engineering and professional skills in a variety of ways at each grade level by infusing the regular curriculum with engineering connections. By implementing this strategy, students, in addition to being exposed to the field of engineering, are expected to be imprinted with some of the professional competencies' qualities or characteristics typical of engineers. Individual "connections" could then be further nurtured at each subsequent grade level or a set of connections could be grouped for presentation at selected grade levels. That is, some connections and subject matter technical content could be addressed in certain grades and not in others. The emphasis of this strategy is to provide students, by the end of 12th grade, a well-rounded exposure to several facets of engineering, thus clarifying what engineers do and demonstrating how they "help" society. It is anticipated that this strategy will not only encourage students to understand the relevance and importance of STEM and non-STEM course content as related to engineering careers but will also minimize students' misperceptions about the profession overall. It focuses on the relationship between engineering concepts and subject matter covered in the current K-12 curriculum. The inclusion of engineering would involve several components such as applications through projects, field trips, and so forth with the objective of enabling students to make the right connections to the relevance of engineering to their present day living.

The question then becomes, how do we best incorporate this idea in the current curriculum? The proposed strategy consists of building an "engineering relevance and application" component into current K-12 curricula. Professor Higgins stated that career imprints bring opportunities as well as constraints depending on how it [career imprinting] is applied (Higgins, 2005; Stark, 2005). In an educational institution setting, opportunities to spark interest in students by using *engineering career imprints* should be sought within the constraints of the regular curriculum, to encourage them to choose engineering as a profession. This strategy should, therefore, not be rigid, but flexible enough to allow variation from state to state as well as from school district to school district, and even from classroom to classroom, if necessary. Some of the variations would depend on schools' structure and the qualifications of teachers. The implementation of this "career imprinting" strategy should ensure that:

- Students acquire and develop grade level, developmentally appropriate, specific knowledge, skills, and *capabilities.*
- Students make the right *connections.* This should involve addressing the social impact of engineering, relevance to daily life, and connections to the various engineering disciplines. These connections will, no doubt, vary from individual to individual depending not only on students' individual personality traits but also on their additional exposure to engineering outside of the school's curriculum.
- Students at all grade levels would be nurtured to develop the *confidence* needed to succeed not only in math, science, and technology-based courses but also in non-STEM courses.
- Students would widen their *cognition* of stereotypes, taken-for-granted assumptions, myths, and so forth that may affect their desire and/or ability to succeed. Examples of some stereotypes include: "Women hate mathematics," "Engineering is boring and has no social relevance," and "Engineers are nerds."

A schematic representation of the strategy, as shown in Figure 7.2, consists of four levels. Emphasis is placed on encouraging students to recognize the relevance of engineering to their own lives, as they now know it. The *first level* involves the establishment of a relationship between engineering concepts and subject matter covered in the current K-12 curriculum. The inclusion of engineering would involve several components such as applications through projects, field trips, and so forth with the objective of enabling students to make the right connections to the relevance of engineering to their present day living. This level will make use of some of ongoing STEM-based subject matter instruction where the application of fundamental principles is already taught. Instruction in non-STEM-based subject matter areas would

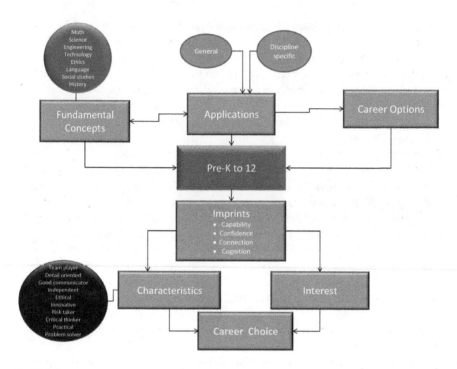

FIGURE 7.2   Schematics of proposed career imprinting strategy.

still be used without much content modification, but opportunities to discuss the relationship of instructional content to engineering and/or technology would be systematically introduced where appropriate.

It is anticipated that both teaching and learning will be enhanced when teachers intentionally make connections between the subject matter being taught and real-world examples that demonstrate how engineering impacts the world in general and the quality of their students' lives in particular. At the end of each grade year, an assessment would be conducted to evaluate the level of capability, connections, confidence, and cognition (4C's) developed. This is the *second level* of the model and can be seen as taking stock of the quality and amount of imprinting already accomplished. It is to be noted that formative assessments can be administered throughout the school year as a way of monitoring students' progress, rather than waiting until the completion of summative evaluations at the conclusion of the school year. The assessment need not be complicated. Grade level appropriateness and the acquisition of evidence to document a student's level of attainment relative to the 4C's would be emphasized. At the *third level*, students who have developed some characteristics and interest in engineering could then receive more detailed instructions such as STEM AP courses, honors classes, and dual-enrollment courses could be used. This level could be broken into two or more streams. In one stream, as already mentioned, students would receive more detailed instruction and preparation for college engineering courses – applicable to upper high school students. Another stream would include further infusion and exposure to hands-on engineering projects, and/

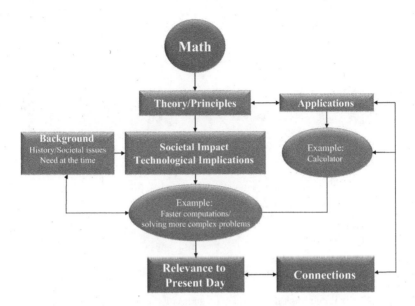

FIGURE 7.3   Example implementation-schematics of strategy for math.

or reinforcement of other grade level specific fundamentals. Students' personal interests in various components of engineering as well as their skills and abilities in fundamental courses would dictate the extent to which they would be involved in self-directed projects and activities related to engineering as a profession. The *fourth and final level* is where students make career choices. Having developed the 4C's through systematic exposure to engineering principles and applications, there is a strong possibility that students would be motivated to pursue careers in engineering. Career imprinting emphasizes connections between and among the subjects which supports the findings in the National Research Council 2014 report (NRC, 2014). The report indicated that a more integrated K-12 STEM education, especially in the context of real world, can make STEM subjects more relevant to students and teachers.

Conceptual strategies for incorporation of career imprinting into the course content for math and social studies are shown in Figures 7.3 and 7.4, respectively. Examples of general classroom and field activities that could be used to reinforce this concept for K-5 and grades 6–12 are presented in Tables 7.1 and 7.2, respectively. Key points that would be highlighted in both examples include:

- What led to the development of such principles?
- How was society impacted then, as well as now?
- Was there any technological/engineering advancement because of this?
- Who were the key players?
- How has or how can this be applied?
- Example application(s)/events specific to each grade level.
- Relevance to present day – making it personal.
- Are the right connections being made? Putting it all together.

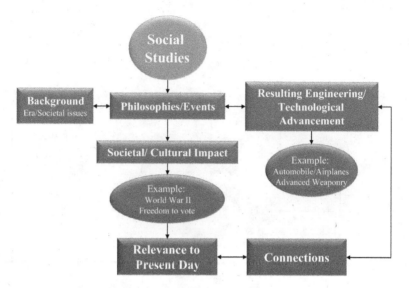

FIGURE 7.4   Example implementation-schematics of strategy for social studies.

TABLE 7.1
Example Activities for Pre-K to Grade 5

Course	Activities
Art	• Design boxes and other containers for special purposes using such mathematical concepts as volume, shape, size, area, etc. • Engage students in learning about types of building construction materials through collage work. • Participate in "I spy" hunts for shapes presented during class. • Provide rebus and/or other simple instructions for students to create designs, structures, and so forth, which require collaborative teamwork that must meet specific "engineering standards" (i.e. must be able to hold five pounds of weight). • Introduce students to the advantages of using different shapes in engineering processes.
Language arts	• Choose literature that emphasizes engineering concepts. • Choose literature that provokes thought about engineering concepts. • Incorporate engineering vocabulary in spelling words and working word walls. • Incorporate engineering vocabulary into students' daily writing and stories. • Create an imagination station that includes dress-up items, engineering tools, pictures of engineering products, etc. for the students to use. • Read fiction and non-fiction books about male and female engineers from around the globe.
Math	• Identify shapes that exist in structures such as buildings and bridges and discuss relationship to engineering. • Incorporate math vocabulary that is specific to engineering into general instructional activities. • Use grade-appropriate calculations to solve math problems relating to various structures or other engineering solutions.

(Continued)

TABLE 7.1 (*Continued*)
Example Activities for Pre-K to Grade 5

Course	Activities
	• Emphasize the importance of understanding and solving perimeter, area, volume, and other algebraic and geometrical concepts as related to the building of roads, waterways, and so forth.
	• Solve daily math problems in a math journal. Show the problems' relevance to engineering.
Music	• Sing songs about engineers and the things they do.
	• Explore "how" instruments work, as students are introduced to them. Then, discuss how engineers design acoustical tiles, spaces, and buildings such as the Opera House in Sydney, Australia.
	• Provide "what if" scenarios for the functioning of instruments, inclusive of engineering-related components.
	• Have students investigate the production of sound with homemade instruments that they share in "show and tell" or "class demonstration" settings, making connections with TV and radio signals as they connect to engineering.
	• Visit music studios and see music production equipment in operation. Encourage students to learn more about sound and electronic engineering.
Physical education	• Engage students in games and other activities where problems must be solved (i.e. rope activities) to develop collaboration, which is a skill regularly used in engineering.
	• Engage students in activities which require team effort and individual effort in order to succeed – to familiarize them with processes used by practicing engineers.
	• Introduce students to charting of their own personal performance. This will introduce students to the skill of maintaining records, which is a major engineering skill.
Science	• Provide opportunities to apply scientific, engineering knowledge. For example, after teaching about simple machines, allow students to construct simple machine houses, cars, or other structures.
	• Take students outside to look for structures that were designed by engineers.
	• Schedule a field trip to a local museum to learn more about engineering products.
	• Encourage students to write in a science journal detailing engineering products and how they are important to their community.
Social studies	• Introduce famous engineers and their accomplishments.
	• Have local engineers visit the school and share their job responsibilities and experiences with the students.
	• Find out who is responsible for the construction of certain structures in the local community and abroad. Engage students in identifying the roles of various engineers who are/were involved in such projects.
	• Have students complete "Then and Now" pictorial histories of various enhancements to quality of life (i.e. washing machines, radio, TV, etc.). Encourage students to make connections between these changes and the work of engineers.
Technology	• Incorporate math websites that encourage practice in geometry and basic algebra.
	• Use the Internet as one resource when researching engineers and engineering products.
	• Help students create presentations about engineers using programs such as PowerPoint, or animated storyboards.
	• Take a virtual tour so students can visit famous structures such as the Eiffel Tower, Statue of Liberty, Leaning Tower of Pisa, San Francisco's Golden Gate Bridge, Gateway Arch in St. Louis, etc.

TABLE 7.2

Example Activities for Grades 6–12

Course	Activities
Art	• Highlight connections between structural stability and architecture from an engineer's perspective. • Functionality of engineering feats. Have students learn about ways in which engineers impact the achievement of such feats by addressing community needs (i.e. building a bridge to connect to small towns). • Role of art in creating housing communities. Encourage students to examine homes that are geometric. • Forces in art (i.e. airbrushing, landscapes, seascapes, etc.). Engage students in discussions about ways in which engineering capitalizes on natural resources (i.e. designing roads through "scenic routes" and soil conservation in flood zones.)
Language arts/English	• Choose literature that emphasizes engineering concepts. • Choose literature that provokes thought about engineering concepts. • Incorporate engineering vocabulary in vocabulary activities. • Incorporate engineering vocabulary into students' daily writing activities. • Incorporate the writing of "user's manuals" into "product activities," since they are a part of engineers' documentation work.
Math	• Incorporate math vocabulary that is specific to engineering (i.e. Word of the Day) into other subjects. • Use grade-appropriate calculations to solve math problems relating to various real-life engineering situations. • Emphasize the importance of understanding and solving perimeter, area, volume, and other algebraic and geometrical concepts. • Require students to work in teams to find suitable solutions to given problems.
Music	• Provide "what if" scenarios for the functioning of instruments to develop problem-solving skills – one used frequently by engineers. • Have students investigate the production of sound with homemade instruments that they share in "class demonstrations" to provide opportunities to share new "concept models" same as in engineering. • Visit music studios and see music production equipment in operation to see what electronics engineers produce. • Engage students in the design and development of audio devices to share music to encourage creative and critical thinking, which is very important in engineering.
Physical education	• Engage students in games and other activities where problems must be solved (i.e. rope activities). • Engage students in activities which require team effort and individual effort in order to succeed. • Engage students in the development and design of equipment that can be used to support P.E. activities.
Science	• Provide opportunities to apply scientific, engineering knowledge in real-world environments. • Schedule a field trip to a local museum to learn more about engineering products. • Encourage students to write in a science journal detailing engineering products and how they are important to their community.

(Continued)

TABLE 7.2 (*Continued*)
Example Activities for Grades 6–12

Course	Activities
Social studies (history, government, political science)	• Introduce famous engineers and their accomplishments. • Have local engineers visit the school and share their job responsibilities and experiences with the students. • Find out who is responsible for the construction of certain structures in the local community and abroad. Visit such sites and write reviews from the perspective of an engineer. • Have students complete "Then and Now" or "Future" pictorial histories of various enhancements to quality of life (i.e. washing machines, radio, TV, etc.) • Involve students in mock court trials to address the impact of engineering-specific implementations. • Involve students in the review of policies and laws related to the implementation of new engineering developments. • Put up bulletin boards that include social issues related to engineering. • Have students interview various types of engineers and discuss issues impacting their work as a result of "political actions." • Have students create a new country and identify "engineers" who are responsible for the "growth and development" of the local regions/communities.
Technology	• Incorporate math websites that encourage practice in geometry, basic algebra, and advanced mathematics. • Use the Internet as one resource when researching engineers and engineering products. • Help students create presentations about engineers using programs such as PowerPoint. • Have students use Excel spreadsheets and other data management tools as a part of regular class work when charts and graphs will be produced. • Take a virtual tour so students can visit famous structures such as the Eiffel Tower, Statue of Liberty, Leaning Tower of Pisa, San Francisco's Golden Gate Bridge, Gateway Arch in St. Louis, etc.

7.3 IMPLEMENTATION CHALLENGES

Infusion of "career imprinting" into day-to-day K-12 classroom instruction, although not a complex process, will certainly have its special challenges.

- One challenge will be increased access to various resources, as teachers will need to have resources available that will enable them to clearly demonstrate connections between what they are teaching and engineering, be they in the form of "what engineers do" or "the impact of engineering" on our world. Such resources might include, but certainly would not be limited to: study trips, guest speakers, special supplies and equipment for projects, audio-visual aids, subscriptions to professional publications, children's and young adults' books about engineering and engineers, attendance at conferences, faculty-engineer exchanges, and so forth.
- Additionally, instructional schedules will need to have the flexibility to deviate from "test-prep or time on task driven" agendas to "capitalizing on

teachable moments" and "student interest-driven" agendas which will afford learners the opportunity to reflect on what they are learning. This will not only make it possible for teachers to put their newly acquired knowledge into the context of what they already know but will also motivate them to think outside of the proverbial "box," thereby setting the stage for more personal internalization of the role of engineers and their impact on society. Through the development of such personal affinity between learners and engineering, it is anticipated that students will be motivated to consider careers in engineering as a whole similar to the experiences students have when "playing doctor" or discussing what they want to be when they "grow up."

- Another challenge, which must not be overlooked, is the school administrator who may not be willing to support the efforts of teachers who are engaging students in more hands-on "fun" activities and projects rather than the common "follow the textbook" path of instruction. Some of the key challenges are discussed in more detail in the following sections.

7.3.1 TRADITIONAL CLASSROOM SCHEDULING

Self-contained classrooms at the elementary school level and departmentalization at the middle and high school levels can be inhibitors to the success of teachers when they attempt to infuse their curriculum with engineering connections. A possible solution for overcoming "structural obstacles" caused by scheduling is to seek opportunities where classwork can be completed via collaboration and students working in teams or presenting information in an interdisciplinary way. This will facilitate the merging of "social studies time" and "math time" at the elementary school level so that rather than instruction being organized into silos, subject matter content and engineering can be connected in ways that will allow both to be addressed simultaneously. For example, having upper elementary students build an operational aqueduct during a lesson on Rome would most certainly require an understanding of mathematical concepts as well as the social studies concept of "providing services to a community." High school students could similarly benefit from having teachers in two different departments work together on a project where, perhaps, English and Chemistry classes were combined (during a common time period) so that a chemistry project, for example, were of the type completed by chemical engineers in a "team setting" and documented in writing. The chemistry teacher could grade the assignment for the fulfillment of the chemistry requirements and the same assignment could then be graded by the English teacher based on the written documentation that was provided by the student. Follow-up assignments, such as a "what if" paper or a "reflection entry" in a journal, could further expand the opportunities for infusing engineering into the subject matter already being covered.

7.3.2 RESOURCES

Money, time, and personnel that can be used to support the development of activities, strategies, and instructional materials which infuse engineering into the existing curriculum may be limited or non-existent. Therefore, teachers are encouraged to consider strategic ways in which infusion can occur by working within their present resource availabilities. For example, rather than a single teacher attempting to develop infusion strategies for all content areas he or she teaches, a group of teachers could work collaboratively to identify and develop common set of infusion strategies. Also, enlisting the help of students, parents, and staff members by setting up "engineering connection" boxes where individuals can submit engineering ideas they have seen, read about, visited, and so forth can generate enthusiasm throughout a school because career imprinting then becomes a "total school effort." Teachers can also reach out to professional organizations such as the American Society for Engineering Education (ASEE), American Society of Civil Engineers (ASCE), and Institute of Electrical and Electronic Engineers (IEEE), some of which have ready-to-use resources. There are also opportunities to collaborate with local colleges to participate in ongoing funded outreach programs. Examples of good resources currently available include the following:

- ASEE P12 program: https://prek-12.asee.org/resources/educators/
- ASEE Go For it: http://www.egfi-k12.org/
- ASCE Dream Big Movie: http://www.dreambigfilm.com/about/
- LinkEngineering: http://linkengineering.org
- Engineering is Elementary: https://eie.org
- ASCEOutreach:http://www.asceville.org/,http://www.asce.org/pre-college_ outreach/
- Engineering by design: www.engineeringbydesign
- Project Lead the Way: www.pltw.org
- Engaging your youth through Engineering: www.maef.net

7.3.3 Professional Staff Development

Most teachers are not prepared to teach engineering due to a lack of understanding of the key concepts (Yasar et al., 2006; Nadelson et al., 2015). Lesson plans developed by teachers with such limited understanding will not be aligned with engineering education. Nadelson et al. conducted a study in which they analyzed 42 lesson plans drawn from 300 STEM lesson plans created by teachers. They found that teachers communicated incomplete understanding of engineering practices and design with limited educational innovation to their students (Nadelson et al., 2016). They also found that "there is a very low likelihood that teachers have the knowledge, skills and motivation needed to effectively implement the innovations." Capobianco and Rupp also indicated from their studies that teachers are developing lesson plans with potentially constrained understanding of engineering (Capobianco and Rupp, 2014).

Unquestionably, infusion of engineering into curriculum content areas will be only as successful as the ability and commitment of the teachers. Therefore, it is important that teachers work together to develop a reflective approach to teaching that includes thinking about the ways in which engineering connects with the subject matter content they teach. It is recommended that teachers regularly dialog with one another about course content and the ways in which engineers and engineering can be incorporated into their lesson plans. Attendance at engineering education conferences/workshops will help teachers become more familiar with the careers of engineers and engineering concepts. Several opportunities for professional development are available through ASEE, NASA (National Aeronautical and Space Administration), NSF, EiE (Engineering is Elementary), and other organizations. For example, the ASEE program specifically trains teachers to become "teachers of engineering" through the P12 program (ASEE, P12). This program provides professional development opportunities that have been made available to teachers to learn how to teach engineering-related coursework through the professional organizations such as ASEE, P12 program (ASEE P12).

Teachers' approach to teaching engineering varies from school to school. Some states including Massachusetts, Oregon, and North Carolina have taken the lead in developing STEM curriculum for pre-k through 12. The Massachusetts Department of Elementary and Secondary Education recently released their report on a STEM curriculum framework with seven guiding principles for effective STEM education. The authors widely subscribe to the guiding principle that states "An effective science and technology/engineering program develops students' ability to apply their knowledge and skills to analyze and explain the world around them" (MA Dept of Education, 2016, Foster, 2009). North Carolina identified four core areas of engineering (engineering habits of mind, engineering design, systems thinking, and problem-solving) that are the basis of instruction (Bottomley, 2013). Brady et al. introduced a pilot project-based engineering course for future K-12 teachers at the California Polytechnic State University. Students were exposed to the engineering design process through problem-solving and developing potential solutions to a design challenge while applying K-12 content (Brady et al., 2016). This is a good approach that could be duplicated at other undergraduate teacher programs.

Engineering has been included in the Next Generation Science Standards; however, the main focus is on science with some engineering applications of science to help students acquire and apply science knowledge. Nonetheless, this will expose more teachers to engineering and help expand the engineering pipeline. Professional development of teachers to address engineering in the classroom could be an expensive venture and may not be feasible for most schools. The framework of career imprinting can be applied within the current K-12 curriculum, so it may be a more readily affordable approach. However, more teachers will still need to acquire a better understanding of engineering concepts and what constitutes good engineering education.

7.3.4 FEDERAL EDUCATION POLICIES

One federal educational policy that could be a challenge is the "No Child Left Behind" (NCLB) or the updated "Every Student Succeeds Act" (ESSA). NCLB included measures to address achievement gaps among traditionally underserved students. Fulfillment of NCLB requirements affords very little, if any, opportunities during class time to infuse engineering into the existing curriculum because "test preparation" was often the priority which may be offered as a piece of resistance. Over time, NCLB's prescriptive requirements became increasingly unworkable for schools and educators. ESSA focuses on a clear goal of fully preparing all students for success in college and careers. A key highlight of ESSA that could support the infusion of STEM into the school curriculum is the requirement to support and grow local innovations that addresses evidence-based and place-based interventions. Teachers may want to consider developing engineering backpacks filled with books, electronic gadgets, educational software, Lego sets, magazines, newspaper articles, and so forth (things that are already available via textbooks, class assignments, etc.) for elementary school students to take home and complete self-selected assignments. Not only will they have more time at home than at school to appreciate the contents of the backpacks, but handouts that list "off-site" resources can be included to provide parents and other community members with the information they need in order to join students in their infusion activities. Middle school and secondary students might enjoy backpacks as well, but they can also be directed to community events and other available opportunities through the development and distribution of a calendar of events which connects engineering to what students are actually doing in class (i.e. a special exhibit on levers and pulleys at a local trade show). Federal education policies are likely to change with a new Administration. However, to date, there have been no current changes that could be assessed as potential challenges.

7.3.5 PARENTAL RESISTANCE

Parents and other school stakeholders may misunderstand the purpose for infusing engineering into regular subject matter content. Some may even vehemently express their opposition. This is to be expected, since today's schools are frequently the subject of exposés where time-on-task and instructional time for students have been frittered away to the point where "real instruction" might appear to be minimal. Teachers and administrators, however, should direct attention to the reasons engineering is considered to be so important: "Engineering builds the critical thinking and design skills that our students need for today's competitive global economy by asking them to apply their math and science knowledge to solve real-world problems" (Valley City State University). Thus, if we want to continue to advance in the future, engineering must remain an integral part of our world today. According to a paper released by the Partnership for 21st Skills, which is sponsored by Cisco Systems,

> Successful learning environments break through the barriers that separate schools from the real world, educators from each other and policymakers from the communities they serve. Yet, many schools continue to reflect the Industrial Age origins with rigid schedules, inflexible facilities, and fixed boundaries between grades, disciplines, and classrooms
>
> *Baker et al. (2007).*

Career imprinting through the infusion of engineering into schools' general curriculum as previously outlined will not only be a step in the right direction toward increasing the number of future engineers who are in the pipeline but will also contribute toward the creation of learning environments that will nurture students positive attitudes toward science, technology, engineering, and mathematics.

7.4 EXAMPLE OF USCGA OUTREACH PROGRAM (2004–2018)

The United States Coast Guard Academy (USCGA) has been very active in attracting and recruiting students into STEM majors for the last two decades. In the spring of 2004, USCGA developed a new Academy Introductory Mission (AIM) program designed to introduce potential cadet candidates to the Academy. The new AIM program was developed by combining some of the best practices of the previous 20 years of summer programs at the Academy. The AIM week-long program was centered on a single intensive STEM-based activity called Coast Guard Academy Robotics on Water (CGAROW). Students were divided into small teams of 5–10. Each team was provided a box of parts that included rigid foam boards, a six-channel radio controller, computer circuit board, motors, servos, small propellers and threaded shafts, miscellaneous metal, and plastic building materials. In addition, each team received a full-scale photo of all the items in their kit for easy identification and inventory. The goal of the program was to have each team design and build a radio-controlled floating robotic craft for use in a competition simulating Coast Guard operations. During the competition, each team had 4 minutes to complete as many Coast Guard mission-oriented tasks as possible by operating their craft in a pool. Throughout the program, cadets and faculty serve as mentors to the participants.

By the fall of 2008, USCGA had five summers of experience with a winning formula for a high energy robotics competition. The program was however hosted on the Academy's campus with limited exposure to the general public. In October 2008, it was proposed to use a modified version of CGAROW for STEM and diversity outreach to middle and high school students. A new outreach version of the CGAROW program (Mobile-CGAROW) was developed from October 2008 through February 2009. Many subtle changes were made to the robotic components. The major change was replacing the original Innovation First Robotics package with a new plug-and-play VEX robotics system. Another important change was the inclusion of a more descriptive handout. The handout provided key hints to ensure success and accelerated the learning process. In addition, an early pre-competition speed test cut down the amount of time required to build the robot from nine to 4 hours. This allowed the program to be completed in under seven hours during a single school day. The program was used as an outreach tool to high school and pilot implementations were done at several high schools in Puerto Rico and in the United States.

The main goals of the Mobile-CGAROW program are to introduce middle- and high-school students to STEM concepts through a hands-on project and to expose them to the missions of the Coast Guard for the purpose of recruitment. The focus of the program is outreach to schools with diverse student populations. A typical one-day program held at the schools involved grouping participants (maximum 30) into teams of 3–5 students. Like the AIM program, but with limited scope, each team is tasked with designing and building a radio-controlled floating robotic craft for use in a competition that simulates Coast Guard operations. The program consists of three phases:

- *General education or familiarization* – The goal of the education phase is to have students familiarize themselves with the program objectives and the operation or functions of the various parts provided.

- *Design and Build* – In the design phase students are required to build a floating robot and modify available components to complete the required tasks in the competition.
- *Competition* – The competition runs for 4 minutes during which teams use their crafts to complete as many Coast Guard mission-oriented tasks as possible in a pool.

Since the successful pilot implementation in March 2009, Mobile-CGAROW has been hosted at several schools throughout the country. The responses from participants were overwhelmingly positive. This resulted in the establishment of outreach partnerships with schools and student exposure to the Coast Guard Operations and STEM careers. After reviewing numerous comments from both students and teachers, it was felt that developing educational or curricular material for CGAROW will enhance the impact of the program on students interested in pursuing STEM careers. Therefore, providing grade-level appropriate educational material prior to an event will better prepare students to effectively compete and apply the STEM principles learned in the classroom. CGAROW was revised to include STEM-based educational materials, standardization of the equipment, an additional mentoring component, and curriculum development.

The Mobile-CGAROW was also included into the annual Engineering Challenge for the 21st Century Teacher enrichment program usually hosted at USCGA. This program is a week-long STEM-focused experience with participants from middle schools, high schools, community colleges, and college outreach programs designed to enhance the effectiveness of teachers in recruiting and retaining STEM students. Working together with a diverse group of schoolteachers and college faculty is a key component to the curricula development. The insight and perspective provided by each member of the group would ensure successful integration. In developing the curriculum component, faculty from USCGA worked with high school teachers to develop a framework for implementation. The goal of the curricula development component is to use the design and competition phases of CGAROW to emphasize the hands-on demonstration of STEM principles and concepts. The educational material was developed within the context of the existing middle and high school STEM curricula. Unfortunately, due to staffing issues and budget cuts, Mobile-CGAROW has been discontinued. Additional information on the CGAROW program including samples of the curriculum material is included in Appendices 7, 8, and 9.

7.5 CLOSING THOUGHTS

The opportunities for K-12 students to be exposed to engineering and its benefits to the world in general are infinite. Every day, students are exposed to engineering from the time they awake in the morning until they go to sleep at night. The quality of their lives, and everyone else's, is significantly improved because of the work performed by various types of engineers. Yet, all too often, students are unable to identify or articulate the impact of engineering on the world in which they live. This is not the case when students are asked to describe what other professionals such as lawyers, teachers, doctors, dentists, law enforcement officers, and others do. Students are "connected" to these careers – not only through the media but also through the exposure that teachers have provided – beginning as early as preschool and kindergarten. By systematically helping K-12 students understand what engineers do, career imprinting can help them begin to more fully comprehend and value the real-world applications. Furthermore, career imprinting will help students learn to appreciate the contributions engineers have made toward improving the quality of lives around the world and instill in them the importance of acquiring the mathematical, science, and "people" skills needed by engineers. By introducing students to the various aspects of engineering, while simultaneously teaching them cognitive skills that focus on creativity, critical thinking, and real-world problem-solving, they can be guided or imprinted to develop strong interests in engineering as a career. Through strategic planning and the creation of real-world connections, students can be inspired to consider engineering as an essential part of everyday living which provides a pathway for the development of creative ideas that can impact the world. Career imprinting has the potential to encourage K-12 students to fully connect learning and living so they more intimately understand why what they are learning really matters. They will then, hopefully, choose to become engineers who are committed to making the world a better place to live, work, and play now as well as for future generations to come.

REFERENCES

American Society of Civil Engineers (ASCE). (2009, April). "ASCE News." Vol. 34, No. 4.

ASEE P12 Program https://prek-12.asee.org/resources/educators/, accessed June 2021.

Baker, D., Yasar-Purzer, S., Kurpius, S., Krause, S. and Roberts, C. (2007). "Infusing design, engineering, and technology into K-12 teachers' practice." *International Journal of Engineering Education*, Vol. 23, No. 5, pp. 884–893.

Bottomley, L. (2013). "Defining engineering in K-12 in North Carolina." *120th ASEE Annual Conference and Exposition*, June 2013, Atlanta, GA.

Brady, P., Chen, J. and Champney, D. (2016) "Future K-12 teacher candidate take on engineering challenges in a project-based learning course." *123th, ASEE Annual Conference and Exposition*, June 2016, New Orleans, LA.

Capobianco, B. M. and Rupp, M. (2014). "STEM teachers' planned and enacted attempts at implementing engineering design-based instruction." *School Science and Mathematics*, Vol. 114, No. 6, pp. 258–270.

Cunningham, C. M. (2009, April). "Engineering is elementary." *National Academy of Engineering, The Bridge*, Vol. 9, No. 3, pp. 11–17.

Engineering is Elementary website. https://eie.org, accessed June 2021.

Florman, S. C. (1996, January). "Spectacles, knotholes, and engineering." *Technology Review*, Vol. 99, No. 1, p. 55.

Foster, J. (2009, April). "The incorporation of technology/engineering concepts into academic standards in Massachusetts: a case study." *The Bridge*, Vol. 9, No. 3, pp. 25–31.

Hester, K. & Cunningham, C. (2007 June), "Engineering is Elementary: An Engineering and Technology Curriculum for Children." Paper presented at 2007 *ASEE Annual Conference & Exposition*, Honolulu, Hawaii, 10.18260/1-2-1469.

Higgins, M. C. and Schein, E. H. (2005). *Career Imprints: Creating Leaders across an Industry*. John Wiley & Sons, Somerset, NJ.

Katehi, L., Greg, P. and Michael, F. (2009, April) "The status and nature of K–12 engineering education in the United States." *National Academy of Engineering, The Bridge*, Vol. 9, No. 3, pp. 5–10.

Massachusetts Department of Elementary and Secondary Education. (2016, April). *Massachusetts Science and Technology/Engineering Curriculum Framework.* (www. doe.mass.edu).

Museum of Boston. (2007, January). *Engineering is Elementary and Technology Lessons for Children (2007).* Curriculum Development: Elementary School Educational Initiatives (video), Boston, MA.

Nadelson, L., Sias, C. and Siefert, A. (2016). "Challenges for integrating engineering into the K-12 curriculum: indicators of K-12 teachers' propensity to adopt innovation." *123rd, ASEE Annual Conference and Exposition*, June 2016, New Orleans, LA.

Nadelson, L. S., Pfiester, J., Callahan, J. Pyke, P., Hay, A. and Emmet, M. (2015). "Who is doing the engineering, the student or teacher? The development and use of a rubric to categorize level of design for the elementary classroom." *Journal of Technology Education* Vol. 26, No. 2, pp. 22–45.

National Academy of Engineering. (2010). *Standards for K-12 Engineering Education?* National Academies Press, Washington, DC.

National Academies of Science, Engineering, and Medicine. (2018). *Indicators for Monitoring Undergraduate STEM Education.* The National Academies Press, Washington, DC. Doi: 10.17226/24943.

National Aeronautic and Space Administration (NASA). website: https://nasa.gpv.offices/education/programs/descriptions/Students-rd.html, accessed June 2016.

National Research Board Council. (2014). *STEM Integration in K-12 Education: Status Prospect, and an Agenda to Research.* The National Academies Press, Washington, DC, Doi: 10.17226/18612.

National Science Board. (2016). *Science and Engineering Indicators 2016.* National Science Foundation (NSB-2016-1), Arlington, VA.

National Science Board. (2018). *Science and Engineering Indicators 2018.* National Science Foundation (NSB-2018-1), Arlington, VA.

Next Generation of Science Standards. (2013). web access: Next Generation Science Standards (nextgenscience.org).

Stark, N. (2005, February 7). "How career imprinting shapes leaders." Working Knowledge: A First Look at Faculty Research. Harvard Business School Working Knowledge Newsletter. http://hbswk.hbs.edu/item4610.html.

University of Massachusetts. "Engineering for K-12 students." http://www.massachusetts.edu/stem/engineering_concept.html, accessed June 2017.

Valley City State University, Museum of Science, National Center for Technological Literacy. "Museum of science's national center for technological literacy and valley city state university to partner in national education initiative, 'Closing the technology and engineering teaching gap.'" http://www.mos.org/nctl/k12initiatives.html.

Yaşar, Ş., Baker, D., Robinson-Kurpius, S., Krause, S. and Roberts, C. (2006). "Development of a survey to assess K-12 teachers' perceptions of engineers and familiarity with teaching design, engineering, and technology." *Journal of Engineering Education*, Vol. 95, No. 3, pp. 205–216.

Appendix 1

Code of Ethics: The American Society of Civil Engineers

PREAMBLE

Members of the American Society of Civil Engineers conduct themselves with integrity and professionalism, and above all else protect and advance the health, safety, and welfare of the public through the practice of Civil Engineering.

Engineers govern their professional careers on the following fundamental principles:

- create safe, resilient, and sustainable infrastructure;
- treat all persons with respect, dignity, and fairness in a manner that fosters equitable participation without regard to personal identity;
- consider the current and anticipated needs of society; and
- utilize their knowledge and skills to enhance the quality of life for humanity.

All members of the American Society of Civil Engineers, regardless of their membership grade or job description, commit to all of the following ethical responsibilities. In the case of a conflict between ethical responsibilities, the five stakeholders are listed in the order of priority. There is no priority of responsibilities within a given stakeholder group with the exception that 1a. takes precedence over all other responsibilities.[1]

CODE OF ETHICS

I. Society

Engineers
  a. first and foremost, protect the health, safety, and welfare of the public;
  b. enhance the quality of life for humanity;
  c. express professional opinions truthfully and only when founded on adequate knowledge and honest conviction;
  d. have zero tolerance for bribery, fraud, and corruption in all forms, and report violations to the proper authorities;
  e. endeavor to be of service in civic affairs;
  f. treat all persons with respect, dignity, and fairness, and reject all forms of discrimination and harassment;

---

[1] This Code does not establish a standard of care, nor should it be interpreted as such. Printed with permission from ASCE

   g. acknowledge the diverse historical, social, and cultural needs of the community, and incorporate these considerations in their work;

   h. consider the capabilities, limitations, and implications of current and emerging technologies when part of their work; and

   i. report misconduct to the appropriate authorities where necessary to protect the health, safety, and welfare of the public.

II. Natural and Built Environment

Engineers

   a. adhere to the principles of sustainable development;

   b. consider and balance societal, environmental, and economic impacts, along with opportunities for improvement, in their work;

   c. mitigate adverse societal, environmental, and economic effects; and

   d. use resources wisely while minimizing resource depletion.

III. Profession

Engineers

   a. uphold the honor, integrity, and dignity of the profession;

   b. practice engineering in compliance with all legal requirements in the jurisdiction of practice;

   c. represent their professional qualifications and experience truthfully;

   d. reject practices of unfair competition;

   e. promote mentorship and knowledge-sharing equitably with current and future engineers;

   f. educate the public on the role of civil engineering in society; and

   g. continue professional development to enhance their technical and nontechnical competencies.

IV. Clients and Employers

Engineers

   a. act as faithful agents of their clients and employers with integrity and professionalism;

   b. make clear to clients and employers any real, potential, or perceived conflicts of interest;

   c. communicate in a timely manner to clients and employers any risks and limitations related to their work;

   d. present clearly and promptly the consequences to clients and employers if their engineering judgment is overruled where health, safety, and welfare of the public may be endangered;

   e. keep clients' and employers' identified proprietary information confidential;

f. perform services only in areas of their competence; and

g. approve, sign, or seal only work products that have been prepared or reviewed by them or under their responsible charge.

V. Peers

Engineers

a. only take credit for professional work they have personally completed;

b. provide attribution for the work of others;

c. foster health and safety in the workplace;

d. promote and exhibit inclusive, equitable, and ethical behavior in all engagements with colleagues;

e. act with honesty and fairness on collaborative work efforts;

f. encourage and enable the education and development of other engineers and prospective members of the profession;

g. supervise equitably and respectfully;

h. comment only in a professional manner on the work, professional reputation, and personal character of other engineers; and

i. report violations of the Code of Ethics to the American Society of Civil Engineers.

Appendix 2
Multistory Building Design Project

To: Engineering Consultants

From: Instructors

Subject: Design of Multistory Reinforced Concrete Building for the Proposed Coast Guard Facility

We are pleased to inform you that you are hereby retained as Structural & Geotechnical Engineering Consultants to analyze and design a multistory reinforced concrete building for the proposed Coast Guard Sector in the *City of Practical Design*. The facility will consist of an administrative multistory building, waterfront access via bulkhead, retaining wall, parking, boat house, and other buildings to support the mission of the Sector. In your recommendation for the various phases of the project, you are required to submit detailed structural analysis and design calculations, detailed sketches, along with preliminary construction drawings, which result in a safe, functional, and economical facility.

This multistory building project will be used in two courses: Geotechnical Engineering Design and Reinforced Concrete Design. Effort will be made by instructors to guide you through the design process as practiced in the civil engineering industry. In the Reinforced Concrete Design course, the structural analysis and design of the multistory building will be completed in three phases as outlined below. Each phase will result is a section of an engineering design report that is due to the client as per the course schedule. In the event of unforeseen circumstances, the client and consultant can agree on a new delivery and completion date with no penalties.

The design phases of this project will be covered in the Reinforced Concrete and Geotechnical Engineering Design courses; details of the building are shown in Figures 1–4.

PROJECT DESCRIPTION

The proposed Coast Guard Sector Design project shall consist of an administrative building, waterfront access via bulkhead, retaining wall, parking, boathouse, and other buildings to support the mission. Some details are provided in this document, additional details for completion of the different geotechnical phases of the project will be provided later.

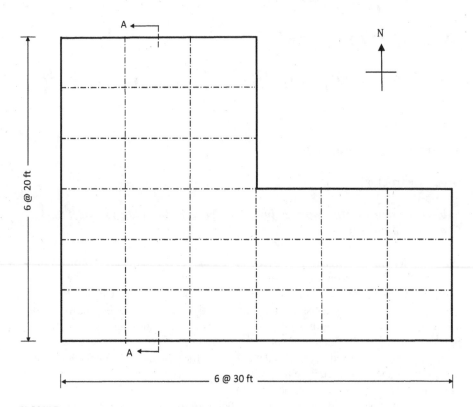

FIGURE 1 Plan view of basic building layout for proposed building. (Dimensions are proposed center to center spacing between columns.)

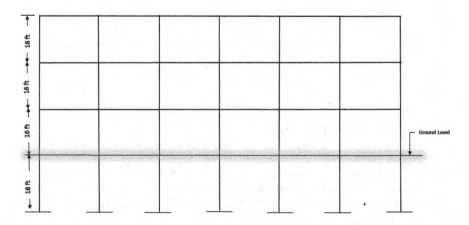

FIGURE 2 Section A-A showing a typical structural frame system for analysis.

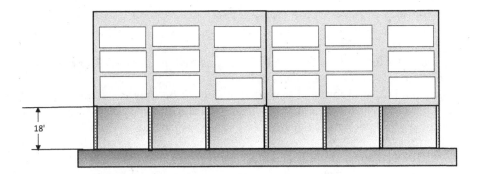

FIGURE 3   Side view W-E direction (looking west).

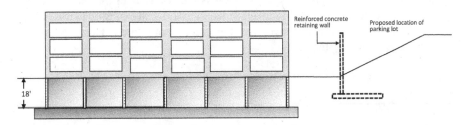

FIGURE 4   Side view S-N direction (looking north).

ADMINISTRATIVE BUILDING DESIGN GENERAL DETAILS

Floor plans and structural framing elevations with column locations for the three-story administrative building have been provided by the architect as shown in Figure 1. The figures provide information regarding the structural configuration and geometry layout. The architect specified a typical 16'-0" floor-to-floor height (dimension from top of structural floor surface to top of structural floor surface noted in building elevation). The fire resistance requirements of the structure shall be:

Roof and Floor beams: 3 hours
Columns (interior & exterior):   3 hours

The unfactored Live and Wind Load requirements for the building follow the governing building code and are estimated as follows:

Roof   Live Load = 40 psf (neglect live load reduction)
            Superimposed DL (mechanical, electrical, ceiling, insulation) = 30 psf
            Superimposed parapet DL at Building Perimeter = 350 lb/ft
            Building Perimeter for façade wall = 60 lb/ft^2
Floor   Live Load = 120 psf (neglect live load reduction)
            Superimposed DL (partitions, ceiling, light, & utilities) = 50 psf
            Building Perimeter for façade wall = 60 lb/ft^2

Wind load on Interior Frame: at Roof level = 12 kips; and at Lower floors = 24 kips

Wind load on Exterior Frame: Multiply estimated load on interior frame load by 0.60.

The loads listed above are SERVICE (unfactored) loads and it does not include the self-weight of concrete members.

The Architect has requested that column sizes to be squares between 12" × 12" and 18" × 18". The main reinforcing bars in the columns shall be at least four and each bar shall not be smaller than #6 and not larger than #11. Use either #3 or #4 ties.

A clear ceiling height of at least 10'-0" is required. The ceiling plenum shall be less than 42" deep which include the false ceiling, lights, telecommunication conduits, and mechanical duct work. The reinforced concrete components of the project shall be designed in compliance with ACI-318. Structural members shall be designed to account for wind load, superimposed loads (dead and live) as well as material self-weight; beam deflection should be limited to L/360, and exposure to the environment or fire. **Use concrete compressive strength (f'_c) of 5,000 psi and ASTM A615 Grade 60 reinforcing bars**.

Reinforced Concrete Design Course

Phase 1- Structural Analysis – Estimate preliminary structural member sizes using practical engineering rules to estimate self-weights. Based on preliminary member sizes and building layout, calculate uniform dead load and live load to be applied on 2D frames (identify critical interior and exterior frames). Wind loads were calculated and provided as point loads at the floor/ roof level. The structural analysis is simplified to only consider three main loads (DL, LL, and W). Perform structural analysis of the building using RISA-2D by considering two factored load combinations (1.2DL + 1.6LL) and (1.2DL + 1.0LL + 1.0W) for each critical frame identified earlier. Submit an engineering report that includes: (1) introduction; (2) preliminary estimates of member sizes; sketches of the selected floor slab, roof joist, beams and columns cross-sections; (3) show applied load combinations on each of the 2D frames (interior and exterior); (4) provide RISA-2D input file data, RISA2D sketch showing node/member numbering system, and RISA sketches showing all member's axial load, shear force, and bending moment results; (5) identify the critical beams at the roof and ground floor using results in step 4. RISA-2D can plot the axial, shear, and bending moment diagrams for all members in a 2D-frame. (6) Summarize critical internal member forces for (a) beams at the roof level, (b) beams at the ground-floor level, and (c) two critical interior and two critical exterior columns at the base of the building. (7) Finally, report the maximum live load deflection of beams predicted by RISA2D at the ground-floor level.

Phase 2- Roof and Floor Design – Update the report introduction and table of content to include all design calculations/sketches of phase 2. Design the roof slab to be a **one-way floor joist** system (slab thickness, stems dimensions and spacing, and all reinforcements) and all lower-level floors to be a **one-way**

solid slab system (slab thickness and all reinforcements). Use RISA-2D results from Phase 1 to design all critical roof/floor beams (L- and T-shapes) that are expected to carry the applied loads in the framing system. All design elements will be designed to resist bending moments and shear forces. Re-submit all the structural analysis (corrected if needed) from phase 1 and all the new phase 2 detailed hand design calculations of typical two flooring systems (roof and ground floors) in a design report. Please submit detailed design calculations for the one-way joist system, beam design at the roof level, one-way solid slab system at the ground-floor level, and beam design at the ground-floor level. Typically, the results of all design calculations are summarized using detailed sketches for all elements which are used in developing construction drawings. These drawings or sketches must show enough details such as dimensions, reinforcement locations with details, and location of various elements in the building so that a contractor will be able to build the flooring systems. Show details of at least two side views of each structural member and then relate each component to the actual location in the building. Examples of typical details are given in Chapter 10 (text) and class handouts.

Phase 3- Column Design, Foundation design, LL Deflection, Structure Point Software, and Final Design Report – Use RISA-2D structural analysis results summarized in Phase 1 to **design all columns** at the basement of the building. Assume the same column size for the entire height of the building but you may vary reinforcement between floors due to change in loading. If RISA-2D output does not show any bending moment at the critical columns, please use an eccentricity of 12 in of the axial loads to accommodate future bending moments. You can use the interaction diagrams to design all the columns and show sketches of column sizes, arrangement of main reinforcement, tie size and spacing's, and column rebar schedule with construction details (show at least two views) of each column. Use the **Structure Point Software** to compare results with your hand design calculations for interior column, exterior column, one interior beam at the first floor, and one-way solid slab. Design two typical **spread footings** to accommodate one interior column and one exterior column. Calculate the maximum **live load deflection** due to a cracked beam section and compare with deflection obtained in Phase 1 (RISA-2D Output). **Prepare and submit a comprehensive engineering design report that includes all three phases**. The report **must be updated to include design calculations and detailed sketches/drawing** of the final designs that can be used to develop construction drawings for this building.

Geotechnical Engineering Design Course

In the Geotechnical Engineering Design course, this project will be completed in phases as outlined in schedule 1 of the contract Agreement and list below. Each phase is due to the client as per the course schedule. Additional details for each phase of the project will be delivered to you as per the timeline agreed upon and documented in the course schedule. In the event of unforeseen circumstances, the client and consultant can agree on a new delivery and completion date with penalties.

Phase 1: Design of a retaining wall include cost estimate
Phase 2: Design of a sheetpile bulkhead
Phase 3: Design of deep pile foundations
Phase 4: Design of a mat foundation/combined footings

A copy of the standard form of Agreement for professional service related to this project is included in the course supplementary handout.

PRELIMINARY FOUNDATION DESIGN DETAILS

The following preliminary details for the foundation design were provided by the structural engineer. These could be later updated based on changes and recommendations from the client and structural engineer. The **unfactored** loads provided are at the foundation level and can be considered to account for all the loading from the superstructure.

Ground surface or grade (ft):   EL. 100
Bottom of the basement slab (ft):   EL. 82
Allowable Bearing Capacity: Calculated in Geotechnical Design Course

Average unfactored uniform load in the building at basement level = 1,500 psf

Appendix 3
Soil Mechanics Technical Paper Guidelines

The technical paper represents 13% of the course grade. It is assigned to strengthen three very important skills; conducting a literature search, writing a technical paper, and public speaking. Your technical research paper should demonstrate that you are developing the necessary skills to become an independent, life-long learner. An assessment rubric will be used to grade your paper and assess your competencies in the following areas:

1. *Information Literacy* – Information literacy is defined by the National Forum on Information Literacy as "The ability to know when there is a need for information, to be able to identify, locate, evaluate, and effectively and responsibly use and share that information for the problem at hand." This includes demonstrating why your topic is important in the context of soil Mechanics and Foundations, and to be able to identify the type of information and extent of research that is appropriate to support your topic. Information literacy also includes an ability to search for and use a variety of valid technical information sources, to quote and paraphrase sources appropriately, and to use proper methods of citation.
2. *Technical (Content) Competency* – The ability to express technical material in a clear and grammatical manner at a level appropriate to your audience.
3. *Writing Skills* – The ability to produce a well-organized and professional research paper as per stated guidelines or format that makes use of appropriate subsections and properly incorporates tables and figures into the text.

PAPER REQUIREMENTS

Organization

The exact sections and subsections used will depend on the nature of each topic. At a minimum the paper should contain the following sections:

a. Abstract
b. *Introduction* – The introduction should start with a general description of the topic.
c. *The Main Research Section* – This section **should be broken down into logical subsections** that flow naturally from a detailed outline you will develop before writing your paper.

   d. *The Conclusion* – This section should tie together the main concepts from the paper and relate them back to the overall topic. As appropriate, suggestions for future research in your topic can be made.

   e. A bibliography that lists all references used in ASCE format or similar format.

The length of the paper should not exceed ten pages without the bibliography. Additional details on the organization and format of the paper are provided on a separate handout. A paper may be longer if your topic warrants the extra length; however, extra length caused by poor writing or poor organization will cost you points.

REFERENCING

All sources must be properly cited in the text and must be listed in a bibliography at the end of the paper using the ASCE journal guidelines (available at http://www.asce.org) or the guidelines distributed in class. All tables and figures must be numbered sequentially, given a meaningful title, introduced in the text by number (i.e., Figure 1 shows ...), and should be incorporated into the body of the paper as soon as possible after being introduced. **A minimum of 4–6 appropriate technical references** should be used from a variety of sources such as technical books, manuals, reports, journal articles, websites, etc.

Grading Rubric for Written Paper (100 points)

Assessment Areas	Exceeds Expectations (90%–100% of Allotted Points)	Meets Expectations (70%–89% of Allotted Points)	Below Expectations (<70% of Allotted Points)
Ability to identify the type and extent of information needed. ___/10	The introduction clearly articulates the relevance of the topic to soil mechanics and/or foundation engineering. The scope and extent of research in the entire paper provides excellent background to support the topic.	The introduction touches on the relevance to soil mechanics and/or foundations. The scope and extent of research in the entire paper provides some good background to support the topic, but is not comprehensive.	There is little to no mention of the tie between the topic and soil mechanics and/or foundations. The type and extent of research is inadequate to provide background for the topic.
Ability to search for and incorporate a variety of appropriate technical information sources. ___/20	At least six appropriate (and credible) technical information sources are used that represent a variety of sources (i.e., technical books, reports, manuals, websites, journal articles, etc.). The sources are varied and comprehensive enough to fulfill the scope of the research without over-reliance on one or two sources.	At least four appropriate (and credible) technical information sources are used that represent at least two types of information sources. The sources generally support the scope of the research and no one source is too heavily relied upon.	There are less than four technical information sources used and/or one or more sources are inappropriate for a technical research paper. One or two sources are too heavily relied upon. The sources do not adequately cover the scope of the research. One or more sources are not credible.
Ability to use sources appropriately, legally, and ethically. ___/10	Sources are properly paraphrased and all direct quotes are in quotation marks. References are cited properly within the text and are appropriately listed in a bibliography at the end of the paper according to the ASCE referencing guidelines.	Sources are properly paraphrased and direct quotes are in quotation marks except for minor issues. References are in the bibliography using the ASCE or similar referencing guidelines.	Sources are not properly paraphrased and/or direct quotes are not properly attributed. Citations are missing from the text and/or are not properly listed in the bibliography according to the ASCE or similar referencing guidelines.

(Continued)

Grading Rubric for Written Paper (100 points) (*Continued*)

Assessment Areas	Exceeds Expectations (90%–100% of Allotted Points)	Meets Expectations (70%–89% of Allotted Points)	Below Expectations (<70% of Allotted Points)
Technical content ___/40	All technical aspects are covered with sufficient details in the context of topic. All important conclusions are clearly stated. Student shows excellent understanding of the relevance of the topic to soil mechanics and foundation engineering.	Most of the important technical details are covered. Only minor details missing. Conclusions regarding major points are stated. Student shows good understanding of the relevance of the topic to soil mechanics and foundation engineering.	Several important technical details are missing or not addressed in sufficient details. Conclusions are missing important aspects of the topic. Student shows lack of understanding of the relevance of the topic to soil mechanics and foundation engineering.
Technical writing skills. ___/10	Writing is clear, concise, and easy to understand. Sentences are grammatically constructed. Passive voice is used. There are no typos or spelling errors. Tables and figures are numbered and properly incorporated into the text after being introduced.	Writing is fairly easy to understand with a few unclear or poorly constructed sentences. There are a few typos or spelling errors. Tables and figures are properly used and incorporated.	Writing is of poor quality. The concepts are poorly explained. There are multiple grammatical errors and/or typos. Tables and figures are not numbered, properly incorporated into the text, and/or introduced.
Ability to format a well-organized paper. ___/10	The paper is well organized as per guidelines or format provided. The introduction explains how the research supports the topic. Subheadings flow logically. The conclusion ties together the main ideas.	The paper is fairly well organized with an introduction, main research section with subheadings, and a conclusion. The conclusion ties together the main ideas. Formatting generally good as per guidelines provided.	The paper lacks good organization. The paper is missing an introduction or conclusion. Formatting does not follow guidelines provided.

Grading Rubric for Presentation: (100 points)

Presenter: _____ Topic: _____

Assessment Areas	Exceeds Expectations (90%–100% of Allotted Points)	Meets Expectations (70%–89% of Allotted Points)	Below Expectations (<70% of Allotted Points)
Organization ___/20	Very clear introduction, conclusion, and consistent sequenced material within body of the presentation. • Information in logical, *interesting* sequence that audience can *easily* follow. • *Engaging* beginning and/or thoughtful ending. Moves smoothly from one idea to the next all of the time.	Clear introduction, conclusion, and consistent sequenced material within body of the presentation. • Information in logical sequence that audience can follow. • Appropriate but not engaging beginning or ending. Moves smoothly from one idea to the next some of the time.	No introduction and/or conclusion and unsequenced material within body of the presentation. • Sequencing is difficult to follow. • Lacks beginning or ending, or inappropriate beginning or ending. Does not move smoothly from one idea to the next.
Technical Content (Coverage of subject matter) ___/40	A variety of supporting materials (e.g., illustrations, statistics, analogies, quotations) make appropriate reference to information or analysis that *significantly* supports the presentation. • Topic covered with sufficient details. • Student demonstrates full knowledge. • Can answer questions fully and accurately.	Supporting materials (e.g., illustrations, statistics, analogies, quotations) make appropriate reference to information or analysis that *generally* supports the presentation. • Most important aspect covered but some minor details missing. • Student demonstrates knowledge of basic concept. • Can answer questions but not fully.	Insufficient supporting materials (e.g., illustrations, statistics, analogies, quotations) make appropriate reference to information or analysis that minimally supports the presentation. • Topic covered with insufficient details. • Student lacks understanding of information. • Cannot answer questions.
Technical Accuracy of Content ___/15	Information is correct and accurate.	Some errors that could distract a knowledgeable listener, but most information accurate.	Information is inaccurate to the extent that presentation cannot be seen as a source of accurate information.

(Continued)

Grading Rubric for Presentation: (100 points) (*Continued*)

Presenter: _____   **Topic:** _____

Category			
Style and Delivery ___/20	Delivery techniques (posture, gesture, eye contact, and vocal expressiveness) make the presentation *compelling*, and speaker appears polished and confident. • Listeners are captivated and focused on ideas presented. • Referred to notes or slides but did not read from them. • Relaxed body language. • Consistent eye contact. • Voice is clear with interesting modulation.	Delivery techniques (posture, gesture, eye contact, and vocal expressiveness) make the presentation *understandable*, and speaker appears tentative. • Listeners can follow presentation, but some distractions. • Reads from notes or slides but also looked away from them. • Slight nervousness. • Occasional unsustained eye contact. • Voice with some inflection.	Delivery techniques (posture, gesture, eye contact, and vocal expressiveness) detract from the understandability of the presentation, and speaker appears uncomfortable. • Listeners are distracted, with difficulty to follow. • Reads from notes or slides and did not look away from them. • Nervous body language. • No effort to make eye contact. • Monotone voice.
Format ___/5	Uniform and consistent fonts and colors. Background and layout of information are appropriate.	Background is appropriate. Good layout, but minor inconsistencies with fonts and colors.	Poor background and inconsistent fonts. Colors, background, and layout inappropriate and very distracting.
Timeline (coordinator's responsibility) [Points Deduction] Comments:	Presentation was within two minutes of allotted time. [*No deductions*]	Presentation was within 2–5 minutes of allotted time. [−1 point deduction/minute]	Presentation was 5 minutes over or under the allotted time. [[−2 points deduction/minute]

Note: If you closely follow the rubric and strive for a level of "Exceeds Expectations" in all assessment areas, you will maximize your grade on the assignment.

Appendix 4
Design Competency Survey

Date:_____

In answering the questions, think of how you would go about (steps you think are necessary) the process of designing and completing a civil engineering project. The questions have been formulated into the steps of the *DRIDS-V* problem-solving framework.

Please indicate your level of competency in the following design components. Select one of the following competency levels for each question.

Competent = I have appropriate knowledge and understand how to apply it in order to complete the task.

Somewhat Competent = I have some knowledge but unsure of appropriateness to complete the task.

Not Competent = I do not have the appropriate level of knowledge to complete the task.

---

1. *Define-Purpose of the Project* – Given the appropriate amount of information, how competent do you feel in defining the problem statement, identifying constraints, and developing the scope of a project?
   Competent   →   Somewhat competent   →   Not competent

2. *Research-Investigate and Gather Relevant Information* – After defining the problem statement, how competent are you in identifying objectives, gathering information from relevant sources such as codes and specifications as well as listing assumptions related to a project? Note that this might result in you refining the problem statement.
   Competent   →   Somewhat competent   →   Not competent

3. *Identify and Decide-Approach to Developing and Selecting Analysis & Design Solutions* – How competent are you in identifying and selecting appropriate analysis, design method(s), and design parameters as well as considering practical constraints and constructability issues related to a project?
   Competent   →   Somewhat competent   →   Not competent

4. *Solve-Apply Appropriate Methods to Analyze and Design Elements or Entire Structure* – After identifying the appropriate design approach, how competent are you in applying it to perform complete analysis and design calculations per standard code(s) and justifying selected analysis approach? (*Understandably, complexity of the process would vary based on the problem.*)
   Competent   →   Somewhat competent   →   Not competent

5. *Verify-Recommendations and Conclusions* – How competent are you in exercising judgment and checking that the results are practical and reasonable and that all aspects of the problem statement and scope were addressed, as well as making recommendations to address pending issues based on your design?
   Competent   →   Somewhat competent   →   Not competent

6. *Design Documents* – How competent do you think you are in preparing detailed design documents including drawings and technical report?
   Competent   →   Somewhat competent   →   Not competent

7. Please provide any additional comments and suggestions.

Appendix 5
Common Grading Rubric (Geotechnical Engineering and Reinforced Concrete Design courses)

	Detailed Description	Exceeds Expectation (>90%)	Meets Expectation (70%–89%)	Below Expectation (<70%)
Define (Purpose of project) _____/5	Define the problem statement, identify constraints and scope of the project. *Bloom's Level 1*	Clear & concise general description of the project scope. Purpose very clearly stated; provides all key details as expressed in the project description. Demonstrates good understanding of problem/assignment.	Concise description of the project scope. Purpose clearly stated; excludes some one or more key details in the project description. Demonstrates fair understanding of problem/assignment.	No general description of the project scope. Purpose unclear, key details excluded; missing important information. Demonstrates poor or no understanding of problem/assignment.
Research (Investigate & gather relevant information) _____/5	Identify & explain objectives, gather information from relevant sources such as codes and specifications, list assumptions. *Bloom's Levels 1, 2, & 3*	All project objectives are listed & clearly explained. Relevant sources of information properly cited. Demonstrates very good understanding of project scope. All of the important assumptions are discussed.	Objectives are identified & listed, but not all are clearly explained. Relevant sources of information properly cited. Demonstrates good understanding of project scope. Most, but not all, of the important assumptions are discussed.	Objectives are not identified & listed. Relevant sources of information not properly cited. Demonstrates poor understanding of project scope. Most of the important assumptions are not discussed.

(Continued)

	Detailed Description	Exceeds Expectation (>90%)	Meets Expectation (70%–89%)	Below Expectation (<70%)
Identify and Decide (Approach to developing & selecting design solutions) _____/10	Identify, select, & formulate appropriate analysis, design method(s), & design parameters. Consider practical constraints and constructability issues. *Bloom's Levels 1, 2, & 3*	Selected suitable analyses & formulated design approaches. Very good engineering judgment and inference; very good understanding of parameters & methods relevant to design.	Identified analyses & formulated design approaches that are mostly suitable (at least one not fully applicable). Good engineering judgment and inference; good understanding of parameters & methods relevant to design.	Suitable analysis & design method not identified or formulated. Poor engineering judgment and inference; little or no understanding of relevant parameters needed for design.
Solve (Apply appropriate methods to analyze and design) _____/50	Conduct detailed analysis and document design calculations. Justify selected design approach. *Bloom's Levels 3 & 4, 5*	Fully applied principles & relevant code with no conceptual or computational errors. Correct calculations & estimation of design parameters, used appropriate correlations. Design is correct & complete. All constraints adequately addressed. Very clear justification of design approach.	Applied principles & relevant code. Few (<5) conceptual or computational errors. Correct estimation of design parameters, used appropriate correlations; accuracy adequate. Most of the key constraints are addressed.	Incorrect calculation of design parameters, used inappropriate correlations. Inability to apply principles & relevant code. Incomplete and/ or inaccurate design. Most of the constraints are not addressed.
Verify (Evaluate Solutions, make Conclusions & recommendations) _____/10	Conduct analysis. Ensure the proposed design adequately addresses all aspects of the problem statement and scope of the project. Exercise judgment & check if design is practical & reasonable. Make recommendations to address pending issues. *Bloom's Levels 5 & 6*	Very good engineering judgment on adequacy of proposed design & decisions is fully supported by code & design requirements. Degree to which solution meets project objectives is well explained. Recommendations address all issues of practicality & constructability.	Good engineering judgment on adequacy of proposed design & decisions is supported by code/ design requirements. Degree to which solution meets project objectives is fairly explained. Recommendations address some but not all issues of practicality & constructability.	Poor engineering judgment on adequacy of proposed design & decisions is not supported by code/ design requirements. Degree to which solution meets project objectives is not explained. Recommendations address some but not all issues of practicality & constructability.

(Continued)

	Detailed Description	Exceeds Expectation (>90%)	Meets Expectation (70%–89%)	Below Expectation (<70%)
Design Drawings _____/10	Illustration design results/recommendations. Drawings & sketches showing relevant dimensions & important details *drawn to scale* in accordance with design calculations. *Bloom's Levels 2 & 3*	Drawings & sketches are complete, accurate in accordance with design calculations & drawn to scale.	Drawings & sketches are drawn to scale in accordance with design calculations, accurate but 10%–20% incomplete.	No drawings or sketches or drawings are inaccurate or drawings not based on design solution or >40% incomplete.
Report Organization _____/10	Professionalism & quality of report. *Bloom's Levels 2 & 3*	Project report is logically organized, well developed, very clear & easy to follow calculations, good transition, good formatting; no grammatical errors.	Project report is logical with good transition and flow, calculations easy to follow, no Few (<5) grammatical errors; little (<3) formatting errors.	Project report is unorganized, poor transition, calculations difficult to follow, several grammatical and formatting errors.

	Detailed Description	Exceeds Expectation (>90%)	Meets Expectation (70%–89%)	Below Expectation (<70%)
Identify and Decide (Approach to developing & selecting design solutions) _____/10	Identify, select, & formulate appropriate analysis, design method(s) & design parameters. Consider practical constraints and constructability issues. *Bloom's Levels 1, 2, & 3*	Selected suitable analyses & formulated design approaches. Very good engineering judgment and inference; very good understanding of parameters & methods relevant to design.	Identified analyses & formulated design approaches that are mostly suitable (at least one not fully applicable). Good engineering judgment and inference; good understanding of parameters & methods relevant to design.	Suitable analysis & design method not identified or formulated. Poor engineering judgment and inference; little or no understanding of relevant parameters needed for design.
Solve (Apply appropriate methods to analyze and design) _____/70	Conduct detailed analysis and document design calculations. Justify selected design approach. *Bloom's Levels 3, 4, & 5*	Fully applied principles & relevant code with no conceptual or computational errors. Correct calculations & estimation of design parameters used appropriate correlations. Design is correct & complete. All constraints adequately addressed. Very clear justification of design approach.	Applied principles & relevant code. Few (<5) conceptual or computational errors. Correct estimation of design parameters used appropriate correlations; accuracy adequate. Most of the key constraints are addressed.	Incorrect calculation of design parameters used inappropriate correlations. Inability to apply principles & relevant code. Incomplete and/or inaccurate design. Most of the constraints are not addressed.

(Continued)

	Detailed Description	Exceeds Expectation (>90%)	Meets Expectation (70%–89%)	Below Expectation (<70%)
Verify (Conclusions & recommendations) _____/10	Conduct analysis. Ensure the proposed design adequately & addresses all aspects of the problem statement and scope of the project. Exercise judgment & check if design is practical & reasonable. Make recommendations to address pending issues. *Bloom's Levels 5 & 6*	Very good engineering judgment on adequacy of proposed design & decisions is fully supported by code & design requirements. Degree to which solution meets project objectives is well explained. Recommendations address all issues of practicality & constructability.	Good engineering judgment on adequacy of proposed design & decisions is supported by code/design requirements. Degree to which solution meets project objectives is fairly explained. Recommendations address some but not all issues of practicality & constructability.	Poor engineering judgment on adequacy of proposed design & decisions is not supported by code/design requirements. Degree to which solution meets project objectives is not explained. Recommendations address some but not all issues of practicality & constructability.
Report Organization _____//10	Professionalism & quality of report. *Bloom's Levels 2 & 3*	Project report is logically organized, well developed, very clear & easy to follow calculations, good transition, good formatting; no grammatical errors.	Project report is logical with good transition and flow, calculations easy to follow, no Few (<5) grammatical errors; little (<3) formatting errors.	Project report is unorganized, poor transition, calculations difficult to follow, several grammatical and formatting errors.

Appendix 6

End of Semester Survey (Geotechnical Engineering Design course)

Section: Instructor: Date:

A. **Rate your level of understanding and ability to apply the basic concepts and principles covered in this course. Please circle one.**
   5: I have a thorough understanding of concepts and principles and can easily apply them.
   4: I have a good understanding of concepts and principles.
   3: I have a fair understanding of concepts and principles or I have some trouble in applying them.
   2: I have a poor understanding of the concepts and principles.
   1: I do not understand the concepts and principles and cannot apply them.

B. **The instructor helped you understand the importance and practical significance of the concepts and design principles discussed in this course. Please select one.**
   5 = strongly agree 4 = agree 3 = neutral 2 = disagree 1 = strongly disagree

C. **Please rate your ability to do the following using the scale below:**
   5 = excellent   4 = very good   3 = good   2 = fair   1 = poor

   ____Select suitable design parameters based on soil profile, construction, economic, and other constraints.
   ____ Analyze the interaction between soil and geotechnical structure in a design problem.
   ____Apply general knowledge to design geotechnical support structures including:
   ____ Retaining structures (sheetpile bulkhead and retaining wall)____ Pile foundations ____Shallow foundations (isolated footing and mat foundation)____ Use engineering codes and design charts

D. **Did the format of the course enhance or diminish your learning experience?**

E. **Which topic or module was not adequately covered?**

F. **Please rate the level of your critical thinking based on the content covered in this course – check all that apply on the first column.**

Check	Critical Thinking Level (Based on 2001 Bloom's Taxonomy)	Description (Action Verbs)
	Remember/Knowledge	I can recall facts and basic concepts (define, duplicate, memorize)
	Understand/Comprehension	I can explain ideas and concepts (classify, describe, identify, discuss)
	Apply/Application	I can use information in new situations (implement, execute, solve, sketch)
	Analyze/Analysis	I can draw connections among ideas (differentiate, organize, compare, examine)
	Evaluate/Evaluation	I can justify a stand or decision (defend, appraise, select, support)
	Create/Synthesis	I can produce new or original work (design, assemble, construct, develop, investigate)

G. Please list your recommendations on improving the course (use back of sheet or additional sheets).

Appendix 7
AROW School Lesson Plan Template

TEACHER LESSON PLAN

Grade Level
Indicate which grade level the lesson is appropriate for.

Objectives
List the learning objectives covered in the lesson.

National Standards Supported
List national standards (science, math, engineering, technology) supported by this lesson.

Required Materials
List of materials needed to complete the lesson.

Step-by-Step Procedures
This section consists of a list of the suggested procedures for implementing the lesson including the duration of the various lesson components.

Assessment
This section contains a listing of the assessments available and included for the lesson.

Remediation
This section contains a description of possible remediation material available for the lesson.

Extension
This section contains a description of a possible extension activity for this lesson.

Supplements
A listing of any additional demo or activities provided to support this lesson.

Sources of Additional Information
List of other resources found on the web or elsewhere that might assist with the teaching of this lesson.

NARRATIVES AND BASIC PRINCIPLES AND CONCEPTS

Narrative (Problem Statement)

Describe a scenario that highlights the problem that is to be addressed using the concepts and principles discussed in this lesson. Tie back to CG missions when appropriate.

Basic Principles and Concepts

This forms the bulk of the lesson material and should be a grade appropriate explanation of the concepts being taught. This section can be used by the teacher as a handout to students or as a guide for preparing their lecture for the day.

Practical Applications

Historical and real-world examples of the principles discussed above and their impact in the real world. When appropriate highlight the STEM fields or disciplines that make use of the principle. How do these principles impact the daily life of students? Where can they see them in action?

DESIGN BRIEF (FOR HANDS-ON EXERCISE)

Instructions for the hands-on activity students will conduct tying back to the lesson. The design brief should include the list of materials and tools required and the deliverables the student must produce. This activity should use as few consumables as possible and should use parts from the AROW kit whenever possible.

Teacher Supplementary Materials:

- Remediation
- Enrichment materials
- Assessment
- Sources of additional information

Appendix 8
Example AROW Lesson Plan on Fundamentals of Buoyancy

TEACHER LESSON PLAN

Topic: Fundamentals of Buoyancy and Stability
Grade Level: 9–12

Metadata:
Naval architecture, Physics, Mathematics, Buoyancy, Stability, Archimedes Principle, Density, Specific Weight, Flotation, Hull design, problems, high school, curriculum, Coast Guard Academy, VEX, AROW

Objectives
- Students will be able to define volume, specific weight, and weight.
- Students will be able to calculate the buoyant force of an object using Archimedes' Principle.
- Students will be able to explain the relationship between buoyant force and weight.
- Students will design and manufacture a model ship's hull.

National Standards Supported
 Standards for Technology Literacy: STL 1.J, STL 1.L, STL 2.R, STL 2.AA, STL 4.I, STL 8.H, STL 9.I, STL 9.K, STL 10.I, STL 10.J, STL 11.P, STL 11.Q, STL 18.J
 NCTM Math Standards
- Understand patterns relations and functions.
- Represent and analyze mathematical situations and structures using algebraic symbols.
- Analyze characteristics and properties of two- and three-dimensional geometric shapes and develop mathematical arguments about geometric relationships.
- Use visualization, spatial reasoning, and geometric modeling to solve problems.
- Apply appropriate techniques, tools, and formulas to determine measurements.

NSTA Science Standards
- Science and Technology Standards
  - Abilities of technological design
  - Understanding about science and technology
- Physical Science
  - Motions and forces

Required Materials
- Aluminum Foil (five sheets per student or student team)
- Container for Water
- Permanent Marker
- Bag of 8-32 Screws
- Foam Block for Constructing Hull (one per student)
- Foam Cutting Tool (Hot Knife or Hacksaw Blade)

Step-by-Step Procedures

1. Using the narrative as an introduction. Students are introduced to the activity that they will be conducting later in the lesson and the role that they will be playing. The focus is on their responsibility as naval architects and the concepts of buoyancy and stability that they will be introduced to in order to fulfill those responsibilities. (Suggested Time: 5 Minutes)
2. Students are introduced to the basic principles and concepts necessary for completion of the activity. The concept of buoyancy and its relationship to Archimedes' Principle is the focus of the lesson. The concepts are explained conceptually and students are instructed on the resulting equations and how they can be applied to solve real world buoyancy problems. Stability is briefly introduced as a concept but the resulting equations are not dealt with at length. Tips for building stable vessels are shared with students and connections to practical real world applications are highlighted whenever possible. You may choose to provide the *Narrative and Basic Principles and Concepts* document to students as a resource and study guide. (Suggested Time: 30 Minutes)
3. The *Design Brief* is distributed to students along with required materials. Students may work independently or in teams on the aluminum boat activity as determined by the teacher but all students should keep copies of the data in their notebooks. Completion of the design brief should be a student directed process. Some students may require additional support in order to complete the Mathematical Modeling portion of the design brief. (Suggested Time: 25 Minutes)
4. The lesson concludes with student AROW teams making a determination as to the hull design they have selected for their vessel submission. One scale model is to be submitted per team along with a rationale as to why that scale model design was chosen. Each team member should include a copy of this rationale and a sketch or picture of their final hull design in their notebooks. (Suggested Time: 35 Minutes)

Assessments
- Final Hull Design Rubric
- Buoyancy and Flotation Questions

Remediation
- Practice Buoyancy Problems

Extension
- Students are to complete a Virtual Physics Lab where they measure the mass of aluminum and iron in and out of water and use this information to calculate specific gravity. Student assessment is done using a printed copy of the student's lab report.
- http://www.polyhedronlearning.com/cengage/lab_18_sim.html

Supplements
- Block/Brick Activity
- Balance Beam – Glove Demo

Sources of Additional Information
- Planet Seed Demonstration of Archimedes' Principle – https://www.planet-seed.com/files/uploadedfiles/Science/Notes/float/en/fands05/FS05_loader.swf
- NASA Video on the Wright Brothers Explaining Lift – http://www.grc.nasa.gov/WWW/Wright/podcast/Podcast_Forces_Lift.m4v
- The Works of Archimedes by T.L. Heath – http://www.archive.org/details/worksofarchimede029517mbp

NARRATIVE AND BASIC PRINCIPLES AND CONCEPTS

In this learning activity, you play the role of a naval architect at Angle Naval Design. Angle Naval Design is a design firm interested in bidding for a United States Coast Guard (USCG) contract to design a new cutter and buoy tender for the fleet. The lead engineer on the project has asked you to submit a ship design for consideration. Your deliverable for this project or the products or services that you are expected to provide consist of a scale model of the final hull design and a rationale for that design.

Buoyancy and stability are crucial concepts if you are to design a vessel or structure that operates as intended on or in the water. An understanding of buoyant force, displacement, and center of gravity are central to properly applying these two concepts.

Engineers use a variety of models to test their designs before moving into production. In this activity, you will first build several small physical models out of aluminum foil to observe the effects of hull shape on buoyancy and stability. You will then proceed to create mathematical models using Archimedes' Principle that allow you to predict the maximum buoyant force acting on a given vessel given the weight of a vessel's superstructure and equipment and the hull's shape and dimensions. Using these models as a guide you will then decide on, sketch, and create a scale model of the hull design you are submitting as your design proposal.

BASIC PRINCIPLES AND CONCEPTS

A piece of steel dropped in water will sink to bottom because it is heavier than water. How then can a ship made out of steel stay afloat in water? Have you ever wondered how a whale floats to the surface and then descends underwater or why an object seems to weigh less in water? These questions can be answered by the principle of buoyancy. Ships and aircrafts all make use of this principle to ensure the safe transportation of goods and people.

Buoyancy and Archimedes' Principle

Eureka! You may have uttered this exclamation before after a surprising discovery but do you know where it originates from or how it is related to buoyancy? (Figure 1) This word is generally attributed to Archimedes of Syracuse a Greek inventor, engineer, mathematician, and physicist who according to folklore became so excited when he stepped into the tub and saw the water level rise that he yelled Eureka and proceeded to run naked through the streets of Syracuse sharing his discovery. Did this really happen? We can't be certain but what we do know is that Archimedes' discoveries, inventions, and observations have had a profound impact on the modern world.

FIGURE 1   Archimedes thoughtful by Fetti (1620).

In Proposition 6 of *On floating bodies,* one of his treatises, Archimedes states that "if a solid lighter than a fluid be forcibly immersed in it, the solid will be driven upwards by a force equal to the difference between its weight and the weight of the fluid displaced."

Archimedes' Principle, the buoyancy theorem that forms the basis for all modern calculations of ship buoyancy, combines and generalizes this proposition and others by Archimedes to state that *the upward force on an immersed (or partially immersed) object is equal to the weight of the fluid displaced by the object.*

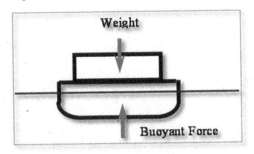

Web_Connection 1: Go to https://www.planetseed.com/files/uploadedfiles/Science/Notes/float/en/fands05/FS05_loader.swf to see Archimedes' Principle demonstrated.

This net upward force is called buoyancy. This force of buoyancy always acts directly upward. The magnitude of this force is directly proportional to the size of the submerged object and can be calculated from Archimedes' principle. The total weight of a vessel must be countered by an equal and opposite buoyancy force. The vertical location of the overall weight determines whether the ship will safely float upright.

$$F_{buoyant} = (\text{displaced volume}) * (\text{specific weight of displaced fluid})$$

$$F_b = v * \gamma \tag{1}$$

where the specific weight of fresh water is 62.4 lb/ft³ or 9.81 kN/m³ and the specific weight of salt water is assumed to be constant at 64.2 lb/ft³ or 10.094 kN/m³.

TEST YOUR KNOWLEDGE!

A solid block of steel is placed in fresh water, will it float or sink? Do you think the density of steel is higher or lower than the density of fresh water? You may ask, what is specific weight? In the case of liquids at the surface of the Earth and at 5°C, specific weight is equal to the density of the fluid in Imperial units. Average density is described as the overall mass of an object over its volume.

The buoyancy force opposes the weight of a floating vessel with an equal and opposite force, but what does it look like? For a box floating in water, the buoyancy

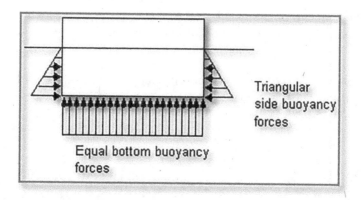

forces press on the bottom and submerged sides of the box. The deeper the box floats in the water the larger the bottom buoyancy forces become. Notice that if the box is floating level, the bottom buoyancy forces are equal, but the buoyancy forces pressing on the submerged sides decrease as we move from the bottom toward the water surface. The side forces appear to be triangular in form. This is because the magnitude of the buoyancy force is directly proportional to the depth of submergence.

We can highlight the implications of Archimedes' Principle for an object immersed or partially immersed in water as the following (Figure 2):

- The weight of the volume of water displaced by any object in water is equal to the buoyancy force (upward force) applied to that object.
- If the weight of the object is less than or equal to the weight of a volume of water equal to the total volume of the object, the object will float.
- If the weight of the object is greater than the weight of a volume of water equal to the total volume of the object, the object will sink.
- The location of the buoyant force is always at the center of the underwater portion of the ship.

FIGURE 2   United States Coast Guard providing assistance to a capsized vessel.

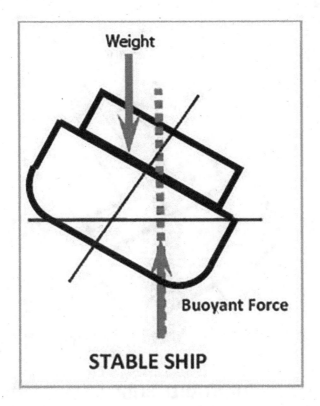

FIGURE 3   Stable ship.

STABILITY

Vessel stability can be defined as the ability to float and to float upright (Figure 3). So, all the forces (weights, buoyancy force) and moments of these forces acting on a vessel must be considered to determine whether a vessel is stable. For a ship, the issue with weight and its location can be summed up in the requirement that a ship should float and it should float upright! The total volume of the hull that can displace water and produce the required buoyant force with a safety margin, called reserve buoyancy, limits the total weight of the ship and its cargo.

Stability is a complex topic which will get a more detailed treatment in another lesson. In order to introduce you to ship stability now however, we must first define two terms that will be crucial to our discussion (Figure 4).

1. *Center of Gravity* – The weight distribution of the ship, its equipment and its cargo determines the ship's center of gravity or the average location of all of the individual weights on the ship.
2. *Center of Buoyancy* – The center of the volume of water displaced by the hull. This location changes as the boat heels (angles) in the water.

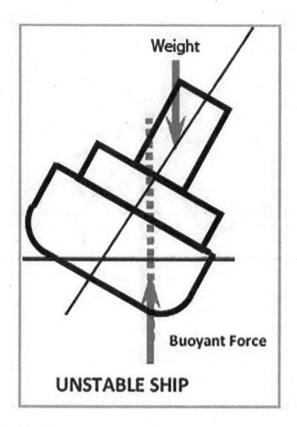

FIGURE 4   Unstable ship.

When a ship heels the center of gravity remains unchanged (assuming the cargo and equipment on board does not shift) but the center of buoyancy changes based on what portion of the ship is now underwater. Buoyant force is now acting on this center of buoyancy with an upwards force while weight is working on the center of gravity with a downwards force. These opposing forces create a torque or a rotational force that is responsible for righting the ship and keeping it vertical in the water.

TEST YOUR KNOWLEDGE!

You are given plans for a vessel, its dimensions and its waterline. Can you calculate its center of buoyancy or center of gravity? What additional information do you need?

For a **STABLE** ship the location of the center of gravity will be such that when the ship *heels* to one side or the other, the righting moment formed by the weight and the buoyant force will cause the ship to rotate back to its original position. For an **UNSTABLE** ship, the location of the center of gravity is such that a heeling moment will occur and the ship will continue to roll until it capsizes.

General Rules for Good Stability

- *Weight distribution affects stability.* Uniformly distributing the weight usually improves stability.
- *The lower the Center of Gravity the better.* However, you may be limited as to how low this can be based on the characteristics of your ship.
- *The wider the better.* A wider ship is generally more stable than a narrow ship. The Buoyant force is able to move farther side to side and may create a larger righting moment. The downside to a wider ship is that it will require more power to travel at a given speed.

TEST YOUR KNOWLEDGE!

If you are on a sailboat heeling to starboard (the right side) and want to lessen the heel where should you go on board? Why? How does this affect the center of gravity?

PRACTICAL APPLICATIONS/MAKE THE CONNECTION

Vessel stability can be considered to be based on weight and moments; a principle that has applications in several branches of engineering. Apart from ships, these principles are at play every time you are on an airplane.

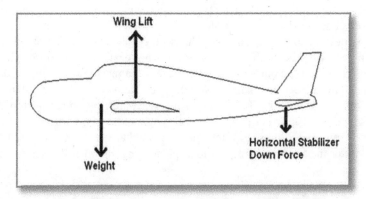

Web_Connection 2: Go to http://www.grc.nasa.gov/WWW/Wright/podcast/ Podcast_Forces_Lift.m4v to watch an explanation of flight.

The aircraft wing produces an upward force to overcome the weight of the aircraft and the horizontal stabilizer produces a downward force. The center of the weight is located forward of the lift force produced by the wing. The downward force produced by the horizontal stabilizer counters the forward pitching moment produced by the weight acting forward of the wing. This is done in order to allow safe recovery in the event of loss of lift force on either the wing or the horizontal stabilizer.

The pilot can increase the lift force on the wing by raising the nose to increase the angle of attack on the wing. The nose is raised by increasing the down force on the horizontal stabilizer. If the overall weight center is too far ahead of the wing lift force, the horizontal stabilizer will not be able to lift the nose. If the overall weight is too far aft, the aircraft will not be able to pitch down for level flight or descent. The importance of knowing the value and location of the aircraft weight prior to take off is obvious. Total weight and weight location limits are carefully established for each aircraft and the pilot is responsible for determining by calculation that the total weight and its location are within those limits prior to each flight.

You've figured out by now that buoyancy and stability are very important to ships and other vessels. The field of engineering that mostly deals with vessel design and stability of ships is *Naval Architecture and Marine Engineering.* Many variables contribute to the design and operation of ships, the most important being the mission or primary operation of the ship. Coast Guard and Navy ships are usually "multi-missioned," so for this discussion, we'll consider the most important design features to be speed and efficiency to accomplish these missions. As a result, the hull shape is narrower and more streamlined. To keep these ships stable, the center of gravity has to be lower so the machinery and tanks are located toward the keel or bottom of the ship and the superstructure is not very high as compared to a commercial ship such as a car carrier, bulk carrier, or cruise ship.

For the commercial sector, speed and efficiency is a design priority, but the main focus of the design is type of cargo it will transport. For example, a car carrier has a wider beam to accommodate the higher superstructure to store cars for transport and still keep the ship stable. A bulk carrier loaded with grains, oil, etc., has a deeper draft which lowers the center of gravity and improves stability as material is loaded onto the ship. Therefore a bulk carrier typically has a narrower beam as compared to a car carrier. The design of a cruise ship is more complicated since a higher super-structure is desired to provide people with scenic views. To ensure stability, you would expect a wider beam, but a wider beam is not desired for a cruise ship since most small vacation-type ports have narrow access channels. As a result, in addition to locating all machinery and tanks toward the keel, ballast tanks and stabilizing fins are used to improve stability. Ballast tanks use sea water to ensure the ship is bal-anced. As fuel is consumed or people move throughout the ship, water can be added or removed from these tanks to counterbalance weight. Stabilizing fins are located below the waterline and work similarly to airplane wings as water flows across them.

MAXIMUM CAPACITIES

7 PERSONS OR 1050 LBS.
1400 LBS. PERSONS, MOTORS, GEAR
130 H. P. MOTOR

THIS BOAT COMPLIES WITH U.S. COAST GUARD
SAFETY STANDARDS IN EFFECT ON THE DATE
OF CERTIFICATION

ABC BOATS
XYZ MANUFACTURING, INC.
ANYWHERE, USA 99999

During the operation of a ship, regardless of the type of ship, somebody is always monitoring the stability of the ship and typically this is the Engineering Officer. Since fuel tanks are usually located near the keel, as fuel is consumed the center of gravity goes up and the stability decreases unless ballast tanks are used. In cold climates, ice is frequently removed from a ship's superstructure because if allowed to accumulate the ice raises the center of gravity. When a Coast Guard ship does migrant interdiction, there is a maximum number of people that can be brought onboard before the ship's stability becomes a concern since the addition of weight on the upper decks of the ship raises the center of gravity. For this same reason, recreational boats are required by the Coast Guard to have a placard indicating the maximum persons and weight to ensure the boat is stable.

DESIGN BRIEF

THE PROBLEM

The Need

The United States Coast Guard (USCG) has conducted a study of its current fleet and missions in order to identify assets that will need to be replaced in the coming 20 years. As a result of that study, it has been determined that it will need to commission 15 new cutters and buoy tenders in the next 5 years. The USCG's Acquisitions Office has asked several firms including yours to submit conceptual designs for review.

Your Job

You are a naval architect at Angle Vessel Design. Your project team has been assigned to work on this project. Your project manager has asked your team to submit a ship design to her for consideration. As a naval architect, your job is to determine the basic size and shape of a vessel, the propulsion requirements, ship structure, weight distribution, and stability.

For this initial phase of the design, your project manager has requested that you submit a scale model of the hull showing the basic size and shape of the vessel. Later you will have the opportunity to further explore propulsion requirements, weight distribution, and stability of the vessel. The scale model your project manager informs you should be designed for a load displacement or total weight of 4.2 lb (18.68 N). This value on a real ship would include the weight of the vessel itself, its passengers, cargo, fuel, and equipment.

THE SOLUTION

The Plan

- Create physical models of ship hulls out of sheets of aluminum foil to allow us to quickly test and visualize new ideas.
- Test the force of buoyancy acting on each of the physical models.
- Familiarize yourselves with the equation for buoyant force and its applications.

- Use mathematical models to calculate the buoyant force of a given ship design.
- Analyze the data collected and use it to make a decision on a hull shape for our conceptual design.

CREATE PHYSICAL MODELS

Earlier we discussed Archimedes' Principle and the idea of buoyant flotation. Buoyant flotation is the product of the weight of the fluid displaced by an object and the specific weight of that fluid.

$$F_{buoyant} = \text{Weight of the Fluid Displaced}$$

$$= (\text{Displaced Volume}) * (\text{Specific Weight of Fluid})$$

$$F_b = v * \gamma$$

Floating: Buoyant Force >= Weight   Sinking: Buoyant Force < Weight

Buoyant force and weight are opposing forces. In the case of a floating object, the buoyant force is greater or equal to the force of weight, resulting in an upward net force.

Take a small screw from your materials and drop it into a container of fresh water.

Now take a sheet of aluminum foil and shape it into the shape of a small boat. Take another small screw the same size as the first but this time, carefully place it into the small boat you just created (Figure 5).

What happens? You will notice that while the first screw sunk to the bottom of the container, the small boat with both designs has been created from the same type and amount of material (in this case a sheet of aluminum foil) one has sunk to the bottom while the other has floated.

The weight of both objects is the same but one sinks and one floats because they have different buoyant forces. How many of the small aluminum balls do you estimate could fit within the small aluminum boat you just floated? This difference in volume or three-dimensional space available within one object accounts for the difference in buoyant force (Figure 6).

Take the remaining three sheets of aluminum foil and carefully shape each into a vessel. This is your opportunity to experiment with different vessel shapes, sizes, and designs (Figure 7).

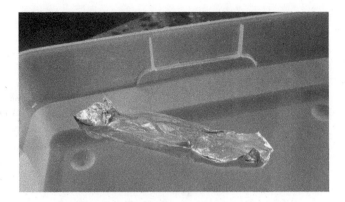

FIGURE 5   Small aluminum foil hull with screw.

FIGURE 6   Aluminum foil hulls.

FIGURE 7   Labeled aluminum hulls.

Label your designs with your name and PM1, PM2, or PM3 using a permanent marker.

Copy the data table below into your engineering or lab notebook.

Test ID	Design Description	Floats Initially	Maximum # of Screws	Stable in the Water
PM 1		Yes/No		Yes/No
PM 2		Yes/No		Yes/No
PM 3		Yes/No		Yes/No

Testing Your Physical Model

Place the first of your physical models into the container of fresh water.

Carefully add screws to your model one at a time. Notice how the waterline on the model changes as additional weight is added and more water is displaced to compensate.

Continue adding screws until your model begins to take on water and sink. Record your data in the data table along with any additional observations (Figures 8 and 9).

Repeat the process for each of your two remaining models.

In your engineering or lab notebook answer the following questions:

1. What hull shapes seemed to have the maximum volume?
2. Which hull shapes had the maximum buoyant force as determined by the number of screws they held?
3. Was this result expected? Why?

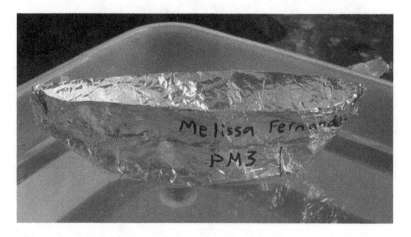

FIGURE 8 Aluminum hull with no screws loaded.

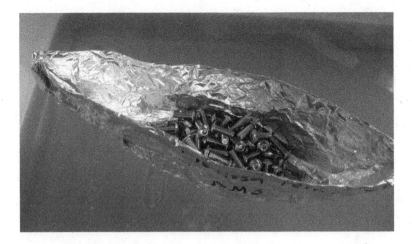

FIGURE 9   Aluminum hull with dozens of screws loaded.

MATHEMATICAL MODELS AND BUOYANT FORCE

At this point, you have explored buoyant forces using physical models and are ready to move on to using mathematical models to explore and predict the behavior of your designs. As we discussed before, the buoyant force acting on an object is equal to the volume of fluid displaced by the object multiplied by the specific weight of the fluid.

$$F_{buoyant} = (\text{Displaced Volume}) * (\text{Specific Weight of Fluid})$$

The specific weight of the fluid varies depending on the fluid but in the case of fresh water, the specific weight is 62.4 lb/ft³ (1,000 kg/m³)

$$F_{buoyant\ of\ fresh\ water} = (\text{Volume of Displaced Water}) * 62.4 \text{ lb/ft}^3$$

With this equation, we can find the maximum buoyant force acting on an object in fresh water given the volume of that object or the ability to calculate its total volume.

As an example, let's take a cube with 2 ft (.609 m) high sides. Using the equation for the volume of a rectangular prism

$$\text{Volume of a Prism} = \text{Length} * \text{Width} * \text{Height}$$

SHOW YOUR WORK!

$$V_{prism} = L * W * H$$

$$V_{prism} = 2\text{ft} * 2\text{ft} * 2\text{ft} = 8 \text{ ft}^3$$

---

we can calculate the total volume of this object.

SHOW YOUR WORK!

$$\text{Displaced Volume} = V_{prism} = 8 \text{ ft}^3$$

$$F_{buoyant\ of\ fresh\ water} = (\text{Displaced Volume}) * 62.4 \text{ lb/ft}^3$$

$$F_{buoyant\ of\ fresh\ water} = 8 \text{ ft}^3 * 62.4 \text{ lb/ft}^3$$

$$F_{buoyant\ of\ fresh\ water} = 499.2 \text{ lb}$$

Because we are calculating the maximum buoyant force, the volume of displaced water is equal to the volume of the object. We substitute the volume of the cube in for the displaced volume and solve to find the maximum buoyant force acting on that object.

Let's now take an example where we know the weight of the object we are placing in the fluid (148 lb) and would instead like to calculate the minimum volume necessary to keep the object afloat. We know that in the case of a floating object, the buoyant force acting on an object is equal to or greater than the weight of the object. Substituting the weight of the object for the buoyant force and calculating for the lowest possible displacement value gives us the following equation:

$$W_{object} = (V_{minimum\ displacement}) * (\text{Specific Weight of Fluid})$$

SHOW YOUR WORK!

In this case

$$F_{buoyant\ of\ fresh\ water} = W_{object} = \text{Load Displacement} = 148 \text{ lb}$$

and

$$F_{buoyant\ of\ fresh\ water} = (\text{Displaced Volume}) * 62.4 \text{ lb/ft}^3$$

therefore

$$\text{Load Displacement} = (V_{minimum\ displacement}) * 62.4 \text{ lb/ft}^3$$

$$148 \text{ lb} = (V_{minimum\ displacement}) * 62.4 \text{ lb/ft}^3$$

$$148 \text{ lb}/62.4 \text{ lb/ft}^3 = (V_{minimum\ displacement}) * 62.4/62.4 \text{ lb/ft}^3$$

$$2.37 \text{ ft}^3 = V_{minimum\ displacement}$$

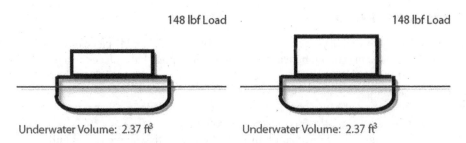

FIGURE 10   Graphic illustrating that the volume of the hull underwater is only dependent on the load displacement not the overall volume of the vessel.

Using this information, we now know that any hull designed to support 148 lb load displacement must have an underwater volume of 2.37 ft³. As we increase the volume of the hull beyond this point we find that the volume of water displaced remains constant as long as the load displacement doesn't change (Figure 10).

APPLIED USE OF THE MATHEMATICAL MODELS

In your engineering or lab notebook sketch two ideas for the shape and size of the hull you are designing. Carefully dimension these sketches. Do not forget to include units in all of your drawings and to label your sketches.

Using Archimedes' Principle and the mathematical models you created earlier, calculate the maximum buoyant force acting on your vessel. If you need assistance calculating the volume of your design, see your teacher for assistance.

Is this buoyant force sufficient to keep a load displacement of 4.2 lb (18.68 N) afloat?

What is the minimum volume of water that must be displaced in order to support the load displacement? Calculate this.

MAKE A DECISION

Using the information gathered from testing your physical and mathematical models it is now time to make a decision as to what hull design you will be submitting for consideration.

Create a sketch of the hull your team has decided to construct. It is important as you create your sketches that you include dimensions that will assist you when marking out what you will need to cut as you shape your foam hull.

Using a pen or pencil mark out the sections of the foam hull you will be cutting. Using the hot knife or hacksaw cut and shape the foam to create the physical model you will be submitting for your deliverable. Safety is of utmost importance during this step. Follow all of your teacher's safety instructions including the wearing of safety glasses as necessary.

Name: _____

Period: _____

BUOYANCY PROBLEMS

1. What is the weight of the water displaced by $1.5'' \times 3.5'' \times 24''$ piece of wood?
2. What is the maximum buoyant force of ten ping-pong balls placed in a mesh bag if the radius of one ping-pong ball is 1 5/8" (ignore the weight of the ball and mesh bag)?
3. If the density of Styrofoam is estimated to be 0.0001 lb/in³, then what is the weight of a $12'' \times 24'' \times 36''$ block of Styrofoam?
4. If the weight of a Styrofoam hull is 0.023 lb what is volume of the hull in inches cubed?
5. If a 3.0 lb water-tight sphere remains half submerged, what is its radius?
6. An undersea archeologist discovers a swivel gun at the ruins of a 17th-century frigate. They plan to raise it by attaching an air-filled bladder. If the cannon weighs 150 lb (undersea weight), what minimum volume of bladder is needed to achieve neutral buoyancy?
7. The USCGC Mackinaw has a total weight of 3,500 long tons (one long ton = 2,240 lb). What is the minimum hull volume required for the vessel to float?
8. A triangular rod of dimensions $1' \times ¼' \times 1/8'$ (length, width, height) with an isosceles triangular cross section with angles measuring ninety and 45° is floating with the 90° angle pointed downward (see diagram). If the weight of the rod is 1/8 lb and the orientation of the rod is constant, what is the draft of the rod?

Name: _____

Period: _____

HULL DESIGN RATIONALE RUBRIC

Task	3 Points	2 Points	1 Points	0 Points
Calculated minimum hull volume based on given load displacement	Minimum hull volume is given with both appropriate units and an explanation of the calculations.	Minimum hull volume is given with either appropriate units or an explanation of the calculations but not both.	Minimum hull volume is given with inappropriate or no units and without an explanation of calculations.	Minimum hull volume is not given or is given incorrectly.
Measured hull volume using appropriate units	Maximum load displacement is given with both appropriate units and an explanation of the calculations.	Maximum load displacement is given with either appropriate units or an explanation of the calculations but not both.	Maximum load displacement is given with inappropriate or no units and without an explanation of calculations.	Maximum load displacement is not given or is given incorrectly.
Calculated theoretical maximum load displacement	Hull volume is given with both appropriate units and an explanation of the calculations.	Hull volume is given with either appropriate units or an explanation of the calculations but not both.	Hull volume is given with inappropriate or no units and without an explanation of calculations.	Hull volume is not given or is given incorrectly.
Vessel stability addressed	Hull stability is addressed including how stability was tested and how it affected design choice	Hull stability is addressed including how stability was tested.	Hull stability is addressed but without details.	No mention of hull stability included in report.
Grammatically correct	No grammar or spelling errors.	1–2 grammar or spelling errors.	3–4 grammar or spelling errors.	5 or more grammar or spelling errors

BLOCK BRICK ACTIVITY

MATERIALS

- Block
- Brick
- Five Gallon Bucket
- Duct Tape

Part I Block

1. Measure the weight, the downward force, of a six-inch length of wooden two by four in pounds.
2. Calculate the volume of the six-inch length of wooden two by four, that is, $1.5'' \times 3.5'' \times 6.0''$ in cubic inches.
3. Calculate the density of the six-inch length of wooden two by four.
4. Calculate the buoyant force of the six-inch length of wooden two by four if it is completely submerged.
   Prediction – Will the block float? (Answer using the language of Archimedes' Principle and buoyant force)
5. Test prediction.

Part II Brick

1. For a given brick measure the volume in cubic inches?
2. Measure the weight of the brick in pounds.
3. Calculate the density of the brick.
4. Calculate the buoyant force of the brick when completely submerged.
5. *Prediction* – Will the brick float? (Answer using the language of Archimedes' Principle and buoyant force)
6. Test prediction.

Part III Block and Brick

1. Attach the wooden block to the ceramic brick using duct tape.
2. Measure the volume of the block/brick system in cubic inches?
3. Measure the weight of the block/brick system in pounds?
4. Calculate the density of the block/brick system?
5. Calculate the buoyant force of the block/brick system if completely submerged?
   Prediction – Will the block/brick system float? (Answer using the language of Archimedes' Principle and buoyant force)
6. Test prediction.

BUOYANCY DEMONSTRATION

This is a wonderful demonstration or challenge problem illustrating buoyancy.

Most chemistry students and teachers will first get this wrong because they are focused on conservation of mass and neglect the result of the buoyant force.

Materials

- Balance Beam or Meter Stick with Pivot Attachment
- Alma-Seltzer Tablet
- Latex Glove
- Counter Weight

PROCEDURE

Fill the finger of latex glove with approximately thirty of water. Put the Alka-seltzer tablet in another finger and hang the glove on balance beam. Level the balance beam using counterweights. Ask the students what will happen if the Alka-seltzers are brought in contact with the water and the glove is allowed to swell in volume.

RESULT

The glove side of the balance beam raises due to the buoyant force equal the weight of the air displaced by the increase in the glove's volume.

BUOYANCY AND FLOTATION QUESTIONS

DEFINITIONS

    1. Archimedes' Principle:
    2. Draft:
    3. Ship Displacement:
    4. Reserve Buoyancy:

THEORY

    1. Write the equation for the buoyant force.
    2. What is the specific weight of fresh water?
    3. What is the specific weight of salt water?
    4. Under what condition will an object (a) sink? (b) float?
    5. What is the weight of a solid object (floating in fresh water) which is 3 ft wide, 6 ft long and 4 ft high with a draft of 3 ft?
       Will that block in Question 5 sink if a 200 lb weight is added?
    6. Why would the side forces acting on a rectangular floating object appear to be triangular in form?
    7. How many long tons is a vessel that weights 23,500 lb?

Appendix 9
Example AROW Lesson Plan on Fundamentals of Torque and Gears

TEACHER LESSON PLAN

Topic: Fundamentals of Torque and Gears
Grade Level: 9–12

Metadata:
Torque, rotational speed, translational speed, gear, gear ratio

Objectives
- Students will be able to define rotational speed, translational speed, torque, and gear ratio.
- Students will be able to use Newton's Third Law and gear ratio to calculate torque and rotational speed.
- Students will calculate translational speed.
- Students will be able to explain the relationship between gear ratio and torque, gear ratio, and rotational speed.

National Standards Supported

Standards for Technological Literacy (STL): STL 1.L, STL 2.X, STL 2.AA, STL 7.G, STL 7.H, STL 8.H, STL 8.K, STL 11.N, STL 11.O, STL 11.P, STL 11.Q

National Science Education Standards (NSES)
Unifying Concepts and Processes – Systems, order, and organization
Unifying Concepts and Processes – Change, constancy, and measurement
Science as Inquiry – Understanding of Scientific Concepts
Science as Inquiry – An appreciation of "how we know" what we know in science.
Science as Inquiry – Skills necessary to become independent inquirers about the natural world.
Physical Science – Motions and forces
Science and Technology – Abilities of technological design
Science and Technology – Understanding about science and technology
History and Nature of Science – Historical perspectives

National Council of Teachers of Mathematics (NCTM)

Numbers and Operations
Understand meanings of operations and how they relate to one another Compute fluently and make reasonable estimates

Algebra
Use mathematical models to represent and understand quantitative relationships

Measurement
Understand measurable attributes of objects and the units, systems, and processes of measurement

Process
Problem Solving Connections Representation

Required Materials
- Foam
- 3-Gear, 60 teeth
- 1-Gear, 36 teeth
- 1-Gear, 12 teeth
- 2-Sprockets, 24 teeth
- Plastic chain (12″ or 30 cm)
- 1–2 Wire Motor 269
- 2-Metal base plate
- 1–25 hole channel bracket
- Assorted screws, bolts, and nuts
- VEX remote control transmitter, signal splitter, and receiver
- Strobe (optional)

Step-by-Step Procedures

1. Using the narrative as an introduction students are introduced to the activity that they will be conducting later in the lesson and the role that they will be playing. The focus is on their responsibility as naval engineers and the concepts of torque and gears that they will be introduced to in order to fulfill those responsibilities. (Suggested Time: 5 Minutes)
2. Students are introduced to the basic principles and concepts necessary for completion of the activity. The concepts of torque and rotational speed and their relationship to gear ratio is the focus of the lesson. The concepts are explained conceptually and students are instructed on the resulting equations and how they can be applied to solve related problems. Torque and rotational speed of gears are briefly introduced as a concept but the resulting equations are not dealt with at length. Tips for meeting speed requirements are shared with students and connections to practical real world applications are highlighted whenever possible. You may choose to provide the Narrative

and Basic Principles and Concepts document to students as a resource and study guide. (Suggested Time: 30 Minutes)

3. The Design Brief is distributed to students along with required materials. Students may work independently or in teams on the activity as determined by the teacher but all students should keep copies of the data in their notebooks. Completion of the design brief should be a student-directed process. Some students may require additional support in order to complete the Mathematical Modeling portion of the design brief. (Suggested Time: 30 Minutes)

4. The lesson concludes with student AROW teams making a determination as to the gear combination (drive train) design they have selected for their vessel submission. One scale model is to be submitted per team along with a rationale as to why that scale model design was chosen. Each team member should include a copy of this rationale and a sketch or picture of their final hull design in their notebooks. (Suggested Time: 25 Minutes)

Assessments
- Final Propulsion or Drive Train Design Rubric
- Torque and Rotational Speed Questions

Remediation
- Practice Gear ratio, rotational speed, and Torque Problems

Extension
- Students can complete a virtual lab in which a beam has to be balanced by changing the location of the pivot point.
- http://www.polyhedronlearning.com/cengage/lab_10_sim.html

Supplements
- Manufacturing of gears
- http://science.discovery.com/videos/how-its-made-gears.html

Sources of Additional Information
- Science discovery website demonstration of Newton Laws of Motion http://science.discovery.com/interactives/literacy/newton/newton.html
- University of Illinois at Urbana-Champaign Department of Physics demonstration of Newton Laws of Motion and other physic principles http://demo.physics.uiuc.edu/lectdemo/
- Neok12 website of on how gears work. Video uses lego to demonstrate how gears change rotational speed and torque. http://www.neok12.com/php/watch.php?v=zX606b6c434d64714f5f6e59&t=Simple-Machines

NARRATIVE AND BASIC PRINCIPLES AND CONCEPTS

In this learning activity, you play the role of a naval engineer at *Marine Propulsion, Inc. Marine Propulsion, Inc* is a design firm interested in bidding for a United States Coast Guard (USCG) contract to design new 41 foot and 47-foot motor-powered response boats for the fleet. The lead engineer on the project has asked you to submit a propeller propulsion system for consideration. The *propulsion* should be based on using the mechanical advantage of a gearing system to attain the speed requirements of the response boats. Your deliverables for this project, or the products or services that you are expected to provide, consist of a scale model of the final propulsion system and a rationale for that design.

Vessels are usually designed to move through the water efficiently. Addressing this involves correctly proportioning and shaping the *hull* (body of vessel), and the size and type of propulsion system includes a source of motive power and a device that converts this power into *thrust*. Newton's laws of motion, mechanical advantage of gears, and propeller action are important concepts in the design of a propulsion system for a vessel to operate on water as intended. An understanding of torque, rotational speed, thrust, and gear ratio is central to the fundamental principles and application of these concepts in the context of moving a vessel in water.

Engineers use a variety of models to test their designs before moving into production. You will be using a battery and motor as the power source and a set of gears and propellers as your propulsion system (or "drive train"). In this activity, you will first build several small physical models using Styrofoam, gears, *sprockets*, chains, motors, and propellers to observe the effects of different gear arrangements on torque, rotational speed, and translational speed. You will then proceed to create models that utilize gear arrangements that allow you to predict the speed of the vessel and torque acting on a vessel given the mass of the vessel and equipment, the hull's shape and dimensions. Using these models as a guide, you will choose a drive train design to submit as your design proposal.

Basic Principles and Concepts

Introduction

Why does a vessel (boat, ship, ferry, yacht, etc.) need propulsion? Unfortunately, we cannot just push off of the beach or dock and expect our boat to continue moving for the rest of our voyage! We must continuously push (or pull!) our vessel to keep it moving. This is because water resists the motion of the boat. Evidence of this *resistance* can be seen in the swirling water behind a moving boat and the waves that are created all around its hull (body of a vessel). One of the goals of hull design is to keep this resistance to a minimum, which therefore reduces the propulsive force (forward force, driving force, or thrust) necessary to keep the boat moving. The propulsion system includes a source of *motive power* (power that produces motion) and a device that converts this power into propulsive force (thrust). The propulsive force is produced by a *mechanical device* called a *propulsor*. The purpose of the propulsor is to push water backwards. The backward movement of water causes an equal and opposite force that pushes the vessel forward. This principle is based on *Newton's third law* that states: "For every action, there is an equal and opposite reaction." The result is that the propulsor provides *thrust* to push the boat forward. There are various types of mechanical propulsors, including *paddlewheels*, *propellers*, and *waterjets*. The most common propulsor on powered vessels is the propeller.

Since you'll be using a set of gears, propellers, and motors to build your drive train, this lesson focuses on gears and how they can be used to change the speed and torque on your vessel. Details on propellers and motors are covered in lesson 2 – *Fundamentals of Marine Propulsion.*

Gears

Gears have been in use for a very long time, since the invention of rotating machinery. Back then, they were made out of wood with cylindrical pegs for cogs. They were very popular in those days because of their multiplying effects...enable people to lift and move heavy loads, etc. Today, gears are mostly made out of metal and plastic and are used for the same reasons as in the olden days. Most gears can be described as circular wheels with teeth. They are usually used to transfer rotational motion and are usually connected to a shaft. Interlocking gears of different diameters can be used to produce a change in rotational speed and rotational force (*torque*). Noncircular gears are designed for special applications such as in textile machines. We will be focusing on circular or spur gears for this lesson.

A device that uses gears that you are familiar with is the bicycle. To ride a multi-speed bicycle, the force from your feet is applied at the pedals which are connected to a gear. As the pedals rotate, torque is transferred from the pedals to the rear wheel by a chain connected to a set of gears. The speed and torque are varied by changing the location of the chain on the rear wheel among the different gears. When climbing up a hill, you'll probably need more torque, so you'd want to move the location of the chain at the rear wheel to a larger size gear. To go faster, you'd want the chain on a smaller gear at the rear wheel. We'll now look at some scientific and mathematical relationships in order to explain how gears work.

> *See how gears are made at:* http://science.discovery.com/videos/how-its-made-gears.html
>
> *See how gears work at:* http://www.neok12.com/php/watch.php?v=zX606b6c4 34d64714f5f6e59&t=Simple-Machines

Torque

In order to tighten or loosen a bolt, it must be turned in a clockwise or anticlockwise direction. How do you get the bolt to turn? Well, you use a wrench to which you apply a force. This applied force act a given distance from the bolt, which in this case is the point of rotation. The applied force multiplied by the distance from the point of rotation (which we will call the *lever arm*) equals the torque.

$$\text{Torque} = \text{Force} \times \text{lever arm} \qquad (1)$$

So, to increase torque, you have two options: increase the force or increase the lever arm. Therefore, for a given force, a longer wrench would make your task of loosening a bolt a lot easier because the torque will be greater.

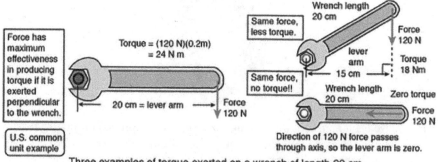

Three examples of torque exerted on a wrench of length 20 cm.

Source http://hyperphysics.phy-astr.gsu.edu/hbase/torq.html

Now let us look at a gear configuration and see how torque comes into play or how it is transferred from one gear to another. *Keep the bicycle example in mind – in cycling, you apply the force with your feet at the pedal and the wheels turn.*

Consider the *free body diagram* of two interlocking gear system shown in Figure 1.

- The radii are of the gears are R_A and R_B.
- When gear *A* (we'll call this the *input gear*) rotates, it generates a force on gear *B* (we'll call this the *output gear*) because they are interlocked at the

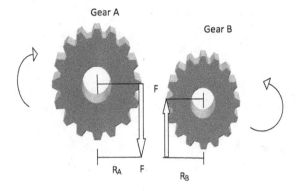

FIGURE 1   Free body diagram of two interlocking gears.

teeth. The force (F) on both gears should be the same magnitude but act in opposite directions (Newton's third law). The gears therefore rotate in opposite directions.

- Applying equation 1, the torque on gear A,

$$T_A = (-F) * R_A \tag{2}$$

- [For simplicity, use force convention with downward as negative and upward positive.]
- Similarly, the torque on gear B,

$$T_B = F * R_B \tag{3}$$

- Dividing T_A by T_B,

$$\frac{T_B}{T_A} = -\frac{R_B}{R_A}, T_B = -T_A\frac{R_B}{R_A} \tag{4}$$

- Or,

$$\text{Output torque} = \text{Input torque} * \text{Gear ratio} \tag{5}$$

- Equation 4 implies that the magnitude of the torque in gear B is proportional to the ratio of the radius of gear B to gear A (gear ratio). Therefore, **if gear B (output gear) is smaller than A, there will be a reduction in torque.** We can then define gear ratio as:

$$\text{The gear ratio (GR)} = \text{Radius of Output gear/Radius of input gear} \tag{6}$$

Gear Ratio and Speed of Rotation

You are familiar with the speed of a car driving along a road. This is actually *translational speed* because the car is moving along a straight line or a curved line. This speed is simply how much distance the car can travel in a given time. It can be calculated by dividing distance by time.

$$\text{Translational Speed} = \text{Distance traveled/time traveled} \qquad (7)$$

Now, think of a spinning bicycle wheel. Since the wheel is spinning about an axis, how fast it rotates about this axis is another type of speed called *rotational speed*. Rotational speed is usually expressed in revolutions per minute (rpm). In one revolution, the wheel rotates or spins 360°.

$$\text{Rotational Speed} = \text{Number or revolutions completed in one minute} \qquad (8)$$

Recall from Figure 1 and equation 1, the ratio R_B/R_A is called the **gear ratio** between two interlocking gears A (input gear) and B (output gear). For a set of gears with the same size teeth, the gear ratio can also be expressed as the ratio of the number of teeth on each gear, that is, N_B/N_A. The gear ratio is important because, for a combination of gears, it determines the magnitude of the torque as well as how fast the shaft connected to the gears rotate. Similar to torque, the rotational speed of two interlocking gears is related to the gear ratio as indicated in equation 9.

$$\text{rpm}_{input} = \text{rpm}_{output} * \text{gear ratio} \qquad (9)$$

Input rotational speed = Output rotational speed*Gear ratio

TEST YOUR KNOWLEDGE ON TORQUE!

1. Gear A pushes on Gear B with .5 pounds of force (2.22 N). What is the force does gear B push back with? Answer in physics terms using Newton's laws.
2. If the medium gear of 1.5″ (0.038 m) diameter is attached to a VEX motor with an output torque of 6.5″*lb (0.73 Nm), then the force on the gear's teeth when meshed with another gear will be?
3. While working on your bicycle, you spin the wheel to check for clearance. The wheel spins nearly without friction. Using the physics terms: torque, radius or lever arm, and force explain why it is better to stop by hand the spinning wheel at the tire rather that grabbing a spoke?

Now consider a propeller attached to a shaft that is connected to the gear configuration shown in Figure 2. The multispeed *motor* is the source of the rotation (hence torque) of the larger gear. This rotation is then transmitted to the smaller gear and shaft connected to it; the first sprocket connected to the same shaft will rotate at the same speed and transfer the rotation to the second sprocket via the chain connecting both sprockets. The second sprocket then transfers the rotation to the propeller through the shaft connected to it. The basic principles for interlocking gears are

FIGURE 2   Gear configuration to increase speed ("Gearing up").

exactly the same for sprockets (gears with thinner teeth) connected by a chain or belt as shown in Figure 2.

Let us use the gear arrangement shown in Figure 2 to determine the torque at the propeller and how fast the propeller would spin given the input torque and input rotational speed of the motor.

Under *very light load* the multispeed motor shown in Figure 2 spins at a maximum of 100 revolutions per minute (rpm); this is the input rotational speed. This speed is only about 2.6 revolutions per second. To increase the speed of the propeller, you need to "gear-up" the shaft RPM as shown in Figure 2. The first (input) gear has *60 teeth* connected to a smaller gear with *34 teeth*. Every time the larger gear spin once, 60 teeth go by and the smaller gear spins 1.765 times (gear ratio = 34/60 or 1:1.765) so that the distance covered by both gears remains the same. Since the sprockets are of the same diameter (or number of teeth), their gear ratio is 1. With the motor spinning at an input rotational speed of 100 rpm, the propeller will therefore spin at 100*1.765*1 = 176.5 rpm. Increasing the output rotational speed of a set of gears is known as "gearing up." You can gear up as many times as you want…but, remember, whenever you increase rotational speed by changing the gear ratio, something happens to torque (see equation 4).

INQUIRY!

What are some of the factors that will affect how fast your boat moves in the water?

What do you think happens to torque as you increase rotational speed by gearing up?

Gearing up increases rotational speed but decreases torque. For the setup shown in Figure 2, the propeller torque is reduced by a factor of the gear ratio (equation 4). If you gear up for twice the output shaft RPM, you get half the output shaft torque. Using the same example in Figure 2, **rotational speed** *increases by a factor of 1.765*, but **torque** is *reduced by a factor of 1.765*. So if your input torque from the motor is 9 lb-in, the output torque at the propeller will be reduced to 5.1 lb-in (9*(32/60)).

If you gear up your propeller shaft by 1:20 or more, the output shaft torque would be so small that the propeller would barely turn. There are other factors that reduce torque such as friction that are not addressed in this lesson. More moving parts and higher chain tension result in greater friction which reduces the torque available to turn a propeller.

PRACTICAL APPLICATION

Mercedes Benz C-class six speed manual transmission showing gear configuration http://auto.howstuffworks.com/transmission-pictures2.htm.

One of the most simple and common practical applications of gears is the bicycle. As previously mentioned, multispeed bicycles are based on moving the chain that connects the pedals to the rear wheel from one size gear to another. Next time you go bike riding, take a closer look at the gear system on your bicycle.

Automobile transmission is also made up a several interlocking gears in order to change from one speed or torque to another. The transmission is designed for a maximum rpm above which the engine might explode. The transmission varies the gear ratio between the engine and the driving wheels as the speed of the car changes. Car engines are also designed to produce a maximum torque at a particular rotational speed or rpm.

Clock gears http://www.plusdeltapartners.com/images/clock_gears.jpg.

If you open up a clock, you'll find lots of gears. They are used to transfer rotational motion of one mechanism to another within the clock.

TEST YOUR KNOWLEDGE ON ROTATIONAL SPEED AND GEAR RATIO!

1. On a flat tabletop place the largest gear (VEX) 2.5″ (0.0635 m) diameter, adjacent to the medium-sized gear of 1.5″ (0.0381 m) diameter. Make sure that the teeth are interlocked. Now rotate the lager gear clockwise.

   What direction does the smaller gear rotate?

   Do the two gears have the same rotational speed and explain?

   Where the teeth of the gears interlock, are the forces same or different and explain?

   What is the gear ratio?

2. A VEX motor turning at 100 rpm is attached to a large gear A (2.5″ diameter) and interlocked with a medium gear B (1.5″ diameter). The medium gear B shares an axial with another large gear C. This large gear C is interlocked with a medium gear D.

   What is the gear ratio from gear A to B?

   What is the gear ratio from gear C to D?

WHAT ARE THE ROTATIONAL SPEEDS OF GEARS A, B, C, AND D? DESIGN BRIEF

THE PROBLEM

The Need

The United States Coast Guard (USCG) has conducted a study of its current fleet and missions in order to identify assets that will need to be replaced over the next 10 years. As a result of that study, it has been determined that the long-range ships (cutters) need to be decommissioned and replace with new long-range cutters. One of these types of new cutters will be the "national security cutter" and is to be equipped with very fast response boats that will be used in drug interdiction and other national security patrols. One of the proposed specifications for the design of the fast response boats is that they attain a maximum speed of 40 *knots*. The USCG's Office of Acquisition has asked several firms including yours to submit conceptual designs for review.

Your Job

You are a naval engineer at *Marine Propulsion, Inc.* Your project team has been assigned to work on this project. Your project manager has asked your team to submit a ship design for consideration. As a naval engineer, your job is to determine the basic size and shape of a vessel, the propulsion requirements, ship structure, weight distribution, and stability.

For this phase of the design, your project manager has requested that you submit a scale model of the propulsion system showing the power source, basic gears and propeller arrangements of the vessel. Your project manager has informed you that the scale model should be designed for a speed of 1.2 *knot* (2.4 ft/s or 0.72 m/s) using a drive train made from a combination of gears, motors, and a power source. In practice, the speed of a real fast boat is specified by the manufacturer and takes into account, the weight of the vessel itself and cargo, type of engine, and propulsion system.

THE SOLUTION

The Plan

- Create physical models of vessel drive train using different size gears and gear combinations to allow us to quickly test and visualize new ideas.
- Determine the speed of each physical model.
- Determine the acceleration and thrust of each physical model.
- Familiarize ourselves with the equations for speed, and thrust and their applications.
- Use mathematical models to calculate the torque, translation, and rotational speed of a given vessel drive train or propulsion design.
- Analyze the data collected and use it to make a decision on a propulsion system for our conceptual design.

BUILD PHYSICAL MODELS

Earlier we discussed Newton's third law of motion. The scientist who postulated this law, Sir Isaac Newton was also a mathematician, astrologer, physicist, and philosopher. He put forth three laws to explain the properties of motion. Since we are dealing with motion or a vessel moving on water, all of Newton's laws are helpful to our study.

Newton's First Law – Every object in a state of uniform motion tends to remain in that state of motion unless an external force is applied to it.
Newton's Second Law – The net force on a body is equal to the product of the body's mass and its acceleration.
Newton's Third Law – For every action, there is an equal and opposite reaction.

Click on the link below to explore Newton's laws of motion. http://science.discovery.com/interactives/literacy/newton/newton.html
Newton's Third Law of motion is of particular interest in this lesson as we look at how torque is transferred from one gear to other.

Torque = Force* Lever Arm

The lever arm is the distance from the force to the center of rotation; so for a circular gear, this would be the radius. We also discussed gear ratio and its effect on torque and rotational speed.

For a two gears combination, the relationships to torque and rotational speed are:

Torque on output gear = Torque on input gear* Gear ratio

Rotational speed on output gear = Rotational speed of input gear/Gear ratio

Take a piece of foam (rectangular or square) and place it in water. Notice that it stays in the same position and does not move if there is no water current. The foam will stay in place if nothing changes around it. Now use your hand to constantly push the foam.

INQUIRY!

Why does the foam now move? What has changed?
Which law of Newton do you think this demonstrates?

Instead of pushing the foam along in the water, we will now look at one way of doing this mechanically. In this exercise, you will investigate several gear arrangement by determining both torque and rotational speed. For the next set of tests, we'll use the foam that has the base plate (see Figure 3) already attached to it – let's call it "the testing platform." Your teacher may already have these testing platforms constructed or will assist you in building one. Instructions on how to construct the testing platform are included at the end of this design brief. Using the testing platform, you will be able to easily attach all of the other components to run your tests.

You'll need a receiver, signal splitter and battery to power your motor as well as a transmitter. These components should be mounted to the base board as shown in Figure 4. Details on wireless communications between the transmitter and receiver are covered in another lesson (lesson 4). For now, we'll just make sure that the receiver matches the transmitter so they can communicate with each other.

Support for Drive
Train Assembly

FIGURE 3   Testing platform with support for drive train assembly.

Receiver (receives signal from
transmitter)

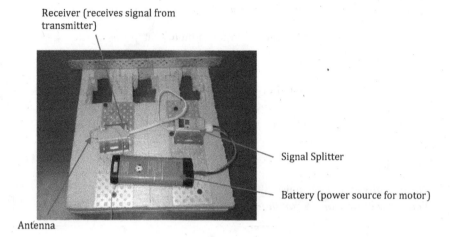

Signal Splitter

Battery (power source for motor)

Antenna

FIGURE 4   Testing platform with battery, splitter & receiver.

FIGURE 5   Tunable strobe light.

First, we want to determine how fast our motor spins under "no load," or "very light load." Typically, this value is provided by the manufacturer. The rotational speed of the multispeed motors you will be using is 100 rpm. As an *optional exercise*, we can measure the rotational speed and compare our value with that from the manufacturer.

Measuring RPM Using a Strobe Light (Optional)

How fast the motor spins can be measured using a tunable strobe light shown in Figure 5. Attach the shaft and propeller to the motor as shown in Figure 6. Mark a

FIGURE 6 Motor with propeller attached.

FIGURE 7 Measuring RPM using a strobe light.

location on one of the propeller blades using a white tape or marker. Connect the motor to one of the channels (use channel 1 or 2) on the signal splitter. Using the transmitter, get the motor running; this will spin the propeller. Turn on the strobe and focus the beam onto the white tape attached to the propeller blade (see Figure 7). Turn the tuning knob on the back of the strobe until the white tape appears stationary or freezes. This is the rotational speed of the shaft in flashes per minute (same as revolutions per minute). Record this value in your lab notebook. Compare this value to that provided by the manufacturer.

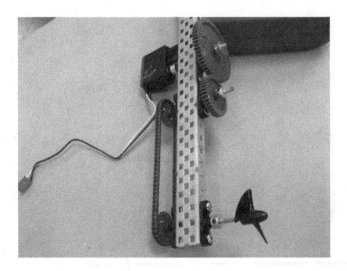

FIGURE 8   General arrangement of gears and sprockets.

Investigating Gear Arrangements

Using the instructions for assembling a drive train presented in lesson 2, the set of gears, sprockets, chains, motor, and propeller provided, assemble the following gear combinations as shown in Figures 5 and 6:

1. One large gear (60 teeth) and one medium gear (36 teeth).
2. One large gear (60 teeth) and one small (12 teeth).
3. One medium and one small.

Make sure you label your designs with your name and GC1, GC2, or GC3 using stickers.

Instructions on how to assemble the drive train and mount it to the testing platform are included as an addendum to this design brief.

The general configuration of the assembled gears and sprockets are as shown in Figures 8 and 9. Note that the two medium sprockets (*24 teeth*) used to transfer the rotation to the propeller are of the same size for each model. Therefore any changes in rotational speed and torque will be due to the gear combination only. Ask your teacher for clarification or assistance if in doubt.

Testing Your Physical Model

Calculate the gear ratio, rotational speed, and torque for each of your gear combination and record them in the data table below (Table 1). Copy Table 1 into your engineering or lab notebook. In your calculation, assume that the input rotational speed and torque of your motor are 100 rpm and 8.6 lb-in, respectively. Use the relationships between gear ratio, torque, and speed to complete this task.

FIGURE 9   Drive trains with three different gear combinations.

TABLE 1
Summary of Data Collected

Gear Combo ID	Design Description	Gear Ratio [-]	Rotational Speed [rpm]	Torque [lb-in]	Translational Speed [ft/s]
GC 1					
GC2					
GC3					

Optional: Measure the output rotational speed of the propeller for each of your gear combination using a strobe light and record them in Table 1. Copy Table 1 into your engineering or lab notebook.

Attach your first physical drive train models onto the support on the testing platform. Do not attach your gears directly to the foam. If in doubt, ask your teacher for help. Place the testing platform with drive train attached into the pool of water and determine the translational speed. Use a stop watch to measure the time your vessel takes to travel a distance (in the pool) predetermined by your teacher. The translational speed can then be calculated using the formula below:

Translational Speed = Distance traveled/time traveled

Record your data in the data table along with any additional observations.
Repeat the process for each of your two remaining models. Plot a graph of:

1. Gear ratio to rotational speed.
2. Gear ratio to translational speed.

In your engineering or lab notebook answer the following questions:

1. Is there a relationship between rotational speed and translational speed? (Another plot of rotational speed versus translational speed may be required.)
2. Which gear combination has the fastest rotational speed? Does this also correspond to the fastest translational speed?
3. Were these results expected? Why?

MATHEMATICAL MODELS AND SPEED

At this point, you have explored several gear combinations and the resulting effect on rotational speed and torque. You can now move on to using mathematical models to explore and predict the behavior of your designs. As we discussed before, the gear ratio influences both torque and rotational speed. Output rotational speed is proportional to the gear ratio and output torque is inversely proportional to gear ratio.

$$\text{Gear Ratio} = \left(\text{Number of teeth on output gear}\right)/\left(\text{Number of teeth on input gear}\right)$$

$$\text{Gear Ratio} = \left(\text{Radius of input gear}\right)/\left(\text{Radius of output gear}\right)$$

$$\text{Rotational speed of gear output gear} = \text{Rotational speed of input gear}/\text{Gear Ratio}$$

As an example, let's say a motor completes 600 revolutions in 4 minutes. What is its rotational speed? Using the definition for rotational speed, we can calculate this value as 150 rpm.

Example

Rotational speed = # of revs in **1** minute = 600/4 = **150 rpm**

It was decided that this speed was too slow for a propeller application. In order to improve the speed, it was proposed to use two interlocking gears-one with 50 teeth and another with 10 teeth. The propeller shaft is attached to the smaller gear. If the input rotational speed of the motor for this setup is 150 rpm, calculate the rotational speed of the propeller.

Example

Given: Input rotational speed of motor = 150 rpm

Number of teeth on interlocking gears

Required: Output rotational speed of propeller

Solution:

Gear Ratio (GR) = # of teeth on output gear/# of teeth on input gear

$= 10/50 = (0.2)$

Or

In ratio form, GR = 10/10: 50/10 = 1:5

Therefore,

The output rotational speed of propeller = input rpm/gear ratio

$= 150/(0.2) = 750$ rpm.

We know that "gearing up" increases rotational speed. Let's now look at an example where we know the minimum expected *output* rotational speed (800 rpm) and would instead like to calculate the gear ratio necessary to attain that speed. So let us calculate the set of gear required to attain a certain rotational speed given the motor output speed (150 rpm).

Example

Given: Input rotational speed of motor = 150 rpm

Expected output rotational speed = 800 rpm

Required: Gear ratio and gear selection

Solution:

Output rotational speed (rpm$_{output}$) = Input rotational speed (rpm$_{input}$) /gear ratio 800 rpm = 150 rpm/gear ratio

Solving for gear ratio,

Gear Ratio = 150/800 = 1: 5.33 (recommend rounding up to at least 1: 6.0)

*Selection of gears:*Use a large gear with **60 teeth as input** connected to **a smaller gear with 10 teeth** as output gear. There are other combinations of gears that would also result in the same gear ratio. Note that the torque will decrease by a factor of 6 in this example. If the output torque is too small, it would be difficult to turn the propeller shaft even though in theory, rpm increases. So there is an upper limit at which "gearing up" becomes ineffective.

Using this approach you now know how to calculate the gear ratio required to meet a specified output rotational speed. In your attempt to design your drive train to meet the translation speed requirement, use the following steps:

1. Note the specified translational speed requirement that must be met.
2. *Estimate Expected Output Rotational Speed* – You should have used the results from the testing of your physical model to establish a trend between translational speed and rotational speed. You should know by now if

translational speed increases or decreases with rotational speed of the propeller. Using this relationship, estimate the rotational speed required. This will be your expected output rotational speed.

3. *Calculate Required Gear Ratio* – Using this output speed and the input rotational speed of the motor (assume 100 rpm), calculate the gear ratio. You can gear up multiple times.

4. *Selection of Gears* – Select an arrangement of gears that will result in the calculated gear ratio.

5. *Assemble Drive Train* – Note that depending on your gear selection, you may need more than one drive train.

APPLIED USE OF THE MATHEMATICAL MODELS

In your engineering or lab notebook sketch two ideas for the gear arrangement for the drive train you are designing. You should already have your hull design completed in lesson 1 (Fundamentals of Buoyancy and Stability). Carefully dimension these sketches. Do not forget to include **units** in all of your drawings and to label your sketches.

Using the concept of gear ratio and the mathematical models you created earlier, estimate the rotational speed of the propeller required to meet the translational speed requirement of 1.2 knots (0.72 m/s or 2.4 ft/s). If you need help completing this task, see your teacher for assistance.

Select the right combination of gears and assemble the drive train.

Attach your drive train to the hull design completed in lesson 1 and measure the translational speed. Record the speed in either metric (m/s) or imperial units (ft/s) and convert to the nautical units of *knots*.

Does the measured speed meet the speed requirements?

Are there any discrepancies between the estimated translational speed based on the rotational speed of the propeller(s) and the measured translational speed?

Note: Depending on how much you gear up, you may need more than one drive train on your vessel to meet the speed requirement. If you use more than one drive train, should they be designed with the same gear ratio? Explain! Record your answer in your notebook.

MAKE A DECISION

Using the information gathered from testing your physical and mathematical models it is now time to make a decision as to what drive train design you will be submitting for consideration.

Create an initial sketch of the drive train attached to the hull including dimensions.

Name: _____ Period: _____

Drive Train Design Rationale Rubric

Task	3 Points	2 Points	1 Points	0 Points
Calculated minimum gear ratio required to meet specific speed	Minimum gear ratio is correctly calculated and reported with explanation of the calculations.	Minimum gear ratio is correctly calculated and reported with slight accuracy in explanation of calculations.	Minimum gear ratio is correctly calculated but reported without an explanation of calculations.	Minimum gear ratio is not given or is incorrectly calculated.
Selected appropriate gears and sprockets to achieve gear ratio	Reported the gears and sprocket sizes and actual gear ratio of the combination.	Reported the gears and sprocket sizes but not the actual gear ratio of the combination.	Reported actual gear ratio but not the gears and sprocket sizes.	Gears and sprocket sizes and actual gear ratio not reported.
Drive train assembly	Sketch or photo of drive train with all components properly labeled.	Sketch or photo of drive train with 1–2 components not labeled.	Sketch or photo of drive train provide without labeling components.	No sketch or photo of drive train provided.
Measured actual translational speed.	Translational speed is correctly calculated and reported in appropriate units with explanation of calculations.	Translational speed is correctly calculated and reported in appropriate units with slight inaccuracy in explanation.	Translational speed is correctly calculated and reported in appropriate units with no explanation of calculations.	Translational speed not reported.
Grammatically correct	No grammar or spelling errors.	1–2 grammar or spelling errors.	3–4 grammar or spelling errors.	5 or more grammar or spelling errors.

SPEED AND TORQUE QUESTIONS

DEFINITIONS

1. Torque:
2. Rotational Speed:
3. Translational speed:
4. Gear ratio:

THEORY

1. What is the equation relating the output torque of two interlocking gears?
2. A large gear (40 teeth) pushes a smaller gear (10 teeth) with a force of 50 N (11.23 lbf). With what force will the smaller gear push back?

3. Rotational speed and torque increase with increase in gear ratio. True or false?

4. The output torque for a gear combination with a gear ratio of 15 connected to a motor is 250 lbf/ins (28.24 N/m). What is the input torque from the motor?

5. A Coast Guard fast response boat travels 5,280 ft in 78 seconds. What is the translational speed in knots? (1 knot = 1.6878 ft/s). If the input rotational speed of the engine is required to maintain this translational speed is 5,000 rpm. What is the output rotational speed of the propeller for a gear ratio of 0.1?

ADDENDUM

Instructions on how to build the testing platform.
Instructions on how to assemble and mount the drive train to the testing platform.

BUILDING THE TESTING PLATFORM

MATERIALS

- 1 – Foam Rectangle Approximately $10\frac{1}{2}'' \times 11\frac{3}{4}''$
- 11 – 1″ Screws
- 1–15×5 Hole Base Plate
- 1–25 Hole Angle
- 1 – Ruler
- 1 – Receiver
- 1 – Signal Splitter
- 1 – Motor Controller
- 1- Multimeter

DIRECTIONS

- Take your foam rectangle and orient it vertically (portrait layout) on your work surface.
- Using the ruler, find and mark the center-line of the block on the top and bottom edge.
- Line up the ruler vertically along the two marks and draw a center-line down the center of your foam piece (Figure 1).
- Using your ruler, draw a horizontal line, 1″ from the bottom of your foam piece.
- On the foam, draw a motor cutaway (Figure 2) centered on the center-line and with the bottom edge touching the horizontal line you have just drawn. If you are only using this testing platform for this activity, you only need to draw and cut out the center motor cutaway. If you will be reusing this testing platform, this would be a good time to draw and cut out all three motor cut outs (Figure 3). Pay attention to the dimensions given. Remember, measure once, cut twice.

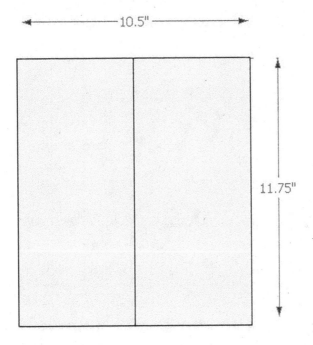

FIGURE 1 Testing platform after drawing a centerline.

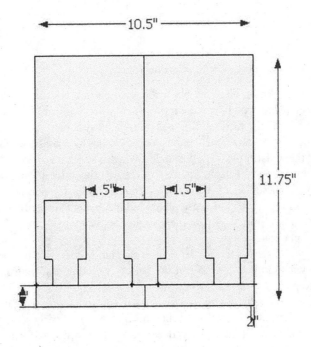

FIGURE 2 Testing platform with three motor cutaways.

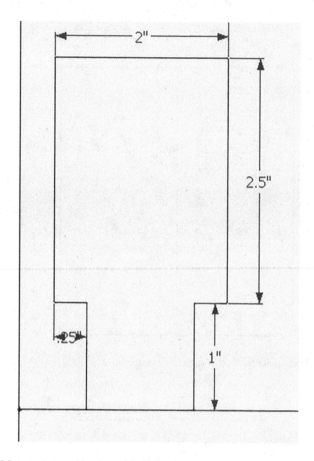

FIGURE 3   Motor cutaway drawing with dimensions.

- Using a hot knife, hacksaw blade, or other cutting instrument cut out each of the cutaways you have drawn on your testing platform. These motor cutaways will be the location for your propulsion systems.
- It is time to place attachment points for the various mechanisms we will be testing on these platforms. Take a 25 hole angle and attach it horizontally as shown in Figure 4. You will attach it to the foam board using three 1" long wood screws as shown. These will serve as your attachment points for your motors.
- Now take two 15×5 hole base plates and place them vertically on the board as shown in Figure 4. Attach these to the foam core using two 1" long wood screws per plate.
- Using #8-32 screws mount the receiver, signal splitter, and motor controller as shown in Figure 5. Connect the battery to the motor controller and zip tie into place.
- Using a multimeter configured to measure DC amperage (see your multimeter's instruction manual for instructions on how to properly set the multimeter) and a stripped motor extender set up your testing platform to measure the current being drawn by the motor. It is important that your meter be

FIGURE 4   Base plate with attachment points.

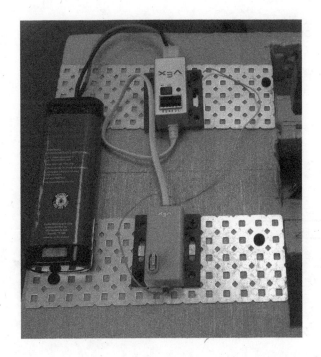

FIGURE 5   The base plate with signal splitter, receiver, and battery connected.

connected in line with the motor if you are measuring current. This means that the red wire of the motor extender cable has been cut and stripped on both sides and each of the multimeter's probes are touching one of the wires. Failure to set up your multimeter correctly inline will cause the fuse to blow on your multimeter.

BUILDING AND MOUNTING THE PROPULSION SUBSYSTEM

MATERIALS

- 1–25 Hole Channel
- 5 – Bearing Flat
- 10 – Screws, #8-32, 1/2"
- 11– Kep Nuts
- 5 – Shaft Collars
- 4 – Spacer, Thin
- 2 – Sprocket, 24 Tooth
- 1 – Threaded Propeller Shaft
- 2- #6-32 Screw, 1/4"
- 1–2 Wire Motor 269
- 1 – Gear, 36 Tooth
- 1 – Gear, 60 Tooth
- 3 – Shaft, 4" (Length can be varied based on availability)

Propulsion Assembly

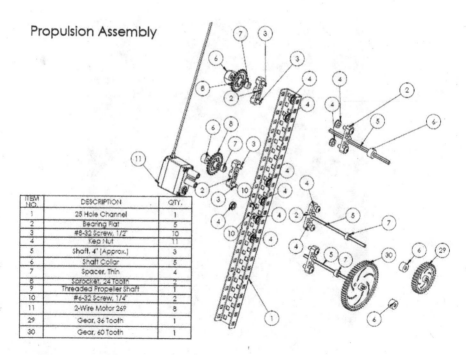

ITEM NO.	DESCRIPTION	QTY.
1	25 Hole Channel	1
2	Bearing Flat	5
3	#8-32 Screw, 1/2"	10
4	Kep Nut	11
5	Shaft, 4" (Approx.)	3
6	Shaft Collar	5
7	Spacer, Thin	4
8	Sprocket, 24 Tooth	2
9	Threaded Propeller Shaft	1
10	#6-32 Screw, 1/4"	2
11	2-Wire Motor 269	8
29	Gear, 36 Tooth	1
30	Gear, 60 Tooth	1

FIGURE 1   Propulsion subsystem assembly.

FIGURE 2   Testing platform in the water.

DIRECTIONS

1. Assemble the propulsion subsystem as indicated below. Pay particular attention to ensuring that only #6-32 motor screws are used on the motors. These motor screws are thinner than the #8-32 screws you normally use to fasten metal in the Vex system.

2. Position the propulsion subsystem in the center motor cut out. Ensure that the threaded screw is pointing toward the stern (back) of the testing platform and is below the water line when the testing platform is placed in the water.

3. Using 2 #8-32 screws attach the propulsion subsystem to the 25 hole angle already mounted to the testing platform.

4. Connect the two-wire motor to the motor controller. The motor controller should be connected into motor input two on the controller.

5. You are now ready to screw on your propeller and begin testing.

Index

Note: **Bold** page numbers refer to tables and *italic* page numbers refer to figures.